The Hobby Farmer's Handbook

TO RAISING
RABBITS
FOR MEAT

Sustainable Rabbit Farming for the Homestead,
from Hutch to Harvest

CANDACE DARNFORTH

Publication Data

Candace Darnforth
Hobby Farmers Handbook to Raising Rabbits for Meat — First edition.
Summary: "Your complete guide to raising rabbits for meat, from selecting breeds and setting up housing to feeding, breeding, and processing. This practical handbook provides step-by-step instructions and expert tips to help hobby farmers successfully manage a sustainable and humane rabbitry"
Provided by publisher.
ISBN: 978-1-961846-12-8
[1. Hobby Farmers Handbook to Raising Rabbits for Meat — Non-Fiction] I. Title.

First paperback edition, 2025

TABLE OF CONTENTS

Introduction

Welcome to the *Hobby Farmer's Handbook to Raising Rabbits for Meat*.

This book is more than a technical manual. It provides a goldmine of information essential for anyone thinking of raising rabbits on a small scale and has everything they need to get started successfully.

The goal of this book is to aid you in everything from selecting ideal breeding stock to designing efficient housing systems; every aspect of rabbitry is meticulously examined. Also, we will discuss how to care for meat rabbits in a way that is good for the environment.

Whether you are planning on raising meat rabbits as a hobby or for profit, you need to learn the fundamental building blocks of establishing and maintaining a rabbitry. This easy-to-read guidebook will equip you with know-how in a few hours of reading. Consider this book your fast track to success.

Let's dive into the world of raising meat rabbits.

CHAPTER 1

Why Rabbits?

According to a recent report, rabbit sales have doubled in the last four-year period and are expected to continue an upward consumption trend. Another report states that rabbit meat is set to see revolutionary growth in the next decade.

Rabbit meat is an excellent source of nutrition due to its high protein content combined with low fat and cholesterol levels. It ranks among the top 10 protein sources and provides more protein per serving than chicken, salmon, and lamb.

Rabbit farming is a rapidly growing market, yet very few agrarians are stepping up to meet the demands of consumers. Getting into meat rabbits is a fantastic opportunity for any new or seasoned farmer eager to learn the lost art of raising rabbits. However, don't get into this business with the goal of making money. While raising rabbits requires relatively low initial costs, space, and labor, it's more of a supplementary income opportunity; even in a good year, many farmers aim to simply cover their costs rather than generate substantial profits.

Raising Rabbits for Meat

Whether you are a newbie or veteran at raising rabbits, the end goal is for meat rabbits to end up in somebody's pot or pan, feeding their family. Rabbits can be braised, baked, fried, fricasseed, stewed, roasted, as well as many other culinary possibilities. The objective is to raise tender, plump dinner rabbits, not cute fluffy bunnies.

In order to attain the goal of tender, plump rabbits, they must be slaughtered when they are still younglings. If the rabbit is too young, it will be tender, but it will not be plump, and if the rabbit is too old, it might be plump but not tender. The ideal age is 10 to 12 weeks for the average meat rabbit.

Compared to other types of farm animals, raising rabbits is relatively easy. Everything about them is light and portable. Unlike other livestock, rabbits are small and docile, which makes them almost effortless to handle. The fact that rabbits and their equipment are so portable makes them the ideal livestock for farmers with physical limitations.

Well-bred rabbits from good breeding stock are often unaggressive, easy to pick up and move from pasture to a cage. Also, rabbits are not heavy. The largest breed of rabbit is the Flemish Giant, which can weigh 22 pounds or more, though their meat-to-bone ratio makes them less ideal for meat production than smaller, more commercial breeds.

Even most of the equipment associated with raising meat rabbits is manageable when relocating your rabbits to a new piece of land for pasturing. The heaviest piece of equipment you will need is a wheelbarrow to haul manure from point A to point B. If you decide to raise your rabbits using a pasture-based method, wire-pasture tractors are light and easy to handle, even more so when compared with portable chicken cages.

Raising meat rabbits is ideal for homesteaders and farmers who do not have access to permanent land tenure. Everything you need for your rabbitry can fit into the back of a large pick-up or into a U-Haul and be moved across the nation. Larger livestock, like dairy cows, require permanent fencing, corrals, a milking parlor, and other infrastructures.

Another feature that makes raising rabbits a solid enterprise is that they require low investment costs to enter the market. Rabbits are small and require little space, even if you decide to use a pasture-based method. Later in this book, we will discuss all the different costs involved in starting a rabbitry, but just to give you a general idea, you can start raising rabbits in your backyard for under $500. A well-managed rabbitry can repay all of the initial investment costs in less than a single year.

Are you still not convinced? As mentioned before, rabbits can be easily transported. Also, you can use containers that you already have, such as a cat carrier or box that can easily fit into the back of your SUV. With time, as your rabbitry expands, you may want to consider investing in a rabbit carrier for $50–$70.

Out of all the types of small livestock, rabbits are one of the easiest to slaughter and dress. For example, chickens require a long list of cumbersome and costly equipment to be processed, such as a kill cone to hold the bird in place, a scalder and a mechanical plucker to remove the feathers, and so much more. Just these three items alone can cost you a couple of thousand dollars. On the other hand, processing rabbits requires as little as a broomstick and a sharp knife (more on butchering and processing rabbits in Chapter 12 of this book).

Another interesting fact about rabbits is their manure. Rabbit manure is packed full of nutrients, such as nitrogen, phosphorus, minerals, potassium, and a long list of micronutrients. According to a study by the Michigan State University Extension, rabbit manure has almost five times as many minerals and nutrients as horse or cow manure and twice as much as chicken manure.

Manure from chicken and pigs (omnivores) is "hot" and needs to be rotted with compost before being used on plants. Hot manure contains too much nitrogen and will burn plants if not properly composted beforehand. Manure from rabbits (herbivores) is "cold," as it has the ideal carbon-nitrogen ratio and can be used on plants without being composted beforehand.

Rabbits produce large quantities of manure. A small rabbitry of 15 does, two bucks, and their offspring can produce more than a ton of manure in a year. Many rabbitries boost their income by selling rabbit manure. Rabbit manure helps improve soil structure. Good soil means better drainage and moisture retention. But gardeners are not the only ones who love rabbit manure. Worms love it, too! For this reason, many rabbitries function as wormeries. Both manure and worms can easily be marketed and sold, thus increasing your capital.

One of the main factors for the increasing popularity of rabbit meat is that it is a lean protein. Rabbit is lower in fat than chicken. For every 85 grams of meat, rabbit has 28 grams of protein, beef (22 grams), turkey (24 grams), pork (23 grams), and chicken (21 grams). Rabbit is an excellent source of iron, with four grams per serving. As for caloric intake, out of all the meats mentioned above, turkey has the least calories, and rabbit meat is the next closest competitor.

All livestock, including rabbits, can have a restorative impact on the environment when raised sustainably. Raising rabbits as a meat source is drawing favorable attention from environmentalists and researchers due to their ecological attributes. Climate change caused by global warming is causing farmers worldwide to look for livestock with a smaller carbon footprint. Most agrarians are struggling with extended droughts and a lack of available arable land. Plus, they are looking for ways to reduce their livestock's carbon footprint. Rabbits require less land, less water, and less energy to grow than conventional livestock.

Finally, as we mentioned earlier, rabbit meat is versatile and delicious. Young rabbit meat is tender and flavorful. It can be prepared in a number of different ways, such as roasted, poached, fried, and braised. In Chapter 12 of this book, you will find some scrumptious recipes.

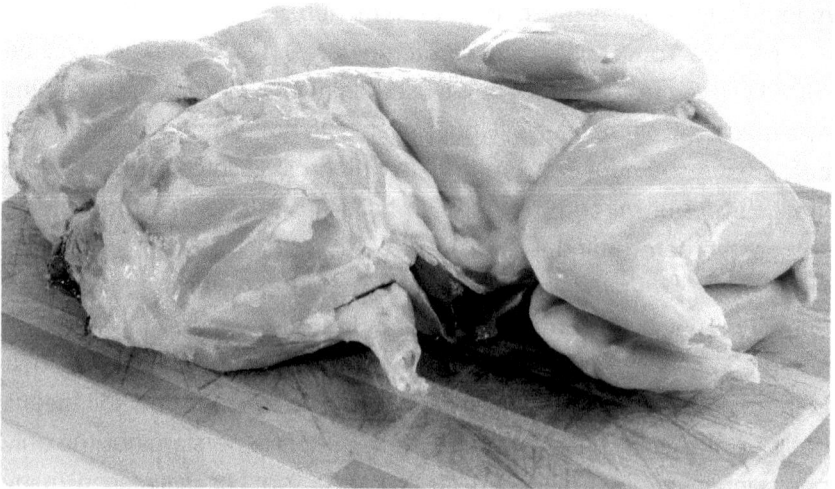

What Is a Rabbitry?

The term "rabbitry" can refer to a small rabbit hutch for a pet rabbit or a larger collection of rabbits that are kept and bred. In this book, when we use the expression, it will refer to a rabbit-raising enterprise for the purpose of producing a meat source.

A rabbit farm, also known as a cunicularium, is one part of cuniculture, the agricultural practice of breeding and raising domestic rabbits as livestock for meat, fur or wool.

In this book, we will discuss in detail how to run a sustainable and profitable rabbitry. We will be focusing on raising meat rabbits, but there are a few words of wisdom to help you take the big step to start your own rabbit farm.

Photo Courtesy of Kelly Hurley, Phillips Farm

Do not go over your maximum occupancy

For obvious reasons, do not breed six litters when you only have the resources for three litters. Additionally, do not bring home four rabbits when you only have the space for two. This may seem like common sense now, but our judgment dulls when we are faced with fuzzy cuteness.

Do not expect to become a millionaire

It does not matter how you work the numbers; it is not very likely that you are going to become a millionaire. Rabbit farming is profitable, and it can generate a healthy income, but you will not get rich from raising rabbits. If you become a millionaire raising rabbits, please let me know your secret!

Check local laws

Be sure to check local laws, especially if you are thinking of starting a rabbitry in your backyard. For example, does your city or homeowners' association have rules on the maximum number of pets? Are rabbits considered to be pets or livestock in your community? While rabbits aren't classified as traditional livestock by the USDA, regulations can vary by location. State and county laws may restrict the age at which rabbits can be sold live to new owners, though specific slaughter age regulations are less common. It's always better to play it safe and check with local authorities beforehand.

Know your breed

Avoid the temptation of choosing more than one breed to start. Do as much research as possible before purchasing your first rabbit. The American Rabbit Breeders Association (ARBA) has established standards for each breed. You can purchase a book, *The Standard of Perfection*, on ARBA's website. Become an expert on your breed.

Prepare for the worst

Rabbits get sick and die, plain and simple. Have a plan of action if your rabbits get sick, such as going to the vet, and set a price limit for each rabbit. If you decide to take your rabbit to the vet, make sure the vet is knowledgeable about meat rabbits. Very few rabbit ailments are curable, so spending $400 to prolong an animal's painful life is no way to treat your animals or spend your money either. Be prepared, study basic ailments and illnesses that can affect your rabbits, have a basic care kit, and remember, not all of your rabbits will live.

Cull hard

Culling means removing certain animals from the breeding process to maintain a healthy, manageable rabbitry. Any rabbit that is sick, weak, or aggressive should be culled to prevent health or behavior issues from spreading through your herd. Kits (baby rabbits) that are weak or sick may not survive and can infect the rest of the litter. Aggressive rabbits are not only difficult to handle but will also pass on undesirable traits, so they should be removed from breeding or moved to a different purpose in the rabbitry.

Research feed options

Research different types of feed options in your area. Pellets are the easiest option and generally the best choice for meat rabbits when you need consistent growth rates to reach target weights efficiently. However, quality matters—poorly manufactured or improperly stored pellets have been known to cause serious health issues in rabbitries due to contamination or mold. Be sure to prioritize quality over quantity. Rabbits are herbivores, not vegetarians. Herbivores eat seeds, leafy greens, grass, and so on. Fruits and veggies can be offered as supplements or treats.

Consider the Rabbit

In European countries, such as France and Spain, rabbit is served for dinner at least once a week. However, if you take a walk down the meat aisle at your local American supermarket, you will find beef, poultry, pork, and occasionally, lamb. In France and Spain, rabbit can be found right beside the poultry section at the grocery store.

Rabbit is a specialty food and often can be pre-ordered or found frozen at European butchers. One of the reasons for this limited availability is that rabbit production in North America typically occurs on smaller farms rather than in large industrial operations, making wide distribution to grocery chain stores less common. Additionally, many North Americans feel intimidated about butchering and cooking a rabbit at home, which affects consumer demand.

So why should North Americans reconsider rabbit as part of their weekly diet?

HEALTH BENEFITS

As mentioned previously, rabbit meat is healthier than beef, pork, lamb, turkey, veal, and chicken. First off, rabbit meat has a higher percentage of protein, the third lowest percentage of fat, and fewer calories per pound. Another health advantage is rabbit contains higher levels of vitamins and minerals like magnesium, potassium, and phosphorus when compared to other conventional meats.

Calories, Protein, & Fat Values for Grilled or Roasted Meat
Per 100 grams (3.5 oz)

	Calories	Protein	Fat (G)
Rabbit	187	27	8
Beef	275	25	20
Pork chops	340	23	24
Lamb	398	22	31
Venison	200	34	6.5
Chicken	190	26	12
Turkey	165	28	6
Duck	330	20	30
Pheasant	250	30	9

Since you will be marketing your rabbit meat, it is important you know some of the health benefits. Besides having a high protein content, it is nearly cholesterol free; therefore, it is a heart-friendly meat option. A bonus is that the sodium level of rabbit meat is relatively low when compared to other meats.

LOWER CARBON FOOTPRINT

The environmental impact of raising rabbits is extremely low because it is sustainable. For example, rabbits produce more than six pounds of meat on the same amount of feed and water as cattle consume to produce only one pound of meat. Also, raising rabbits requires little space, especially when compared to cattle, goats, or pigs, which means less impact on energy resources. Rabbits are foragers, meaning they thrive on a grain-free diet of hay, compost scraps, and grass, which is more affordable, sustainable, and naturally available.

A DELICIOUS OPTION

Besides rabbit meat's health benefits, it is a versatile and delicious option. You can roast, braise, and fry rabbit meat. Butchers will happily break the meat down for customers. If you are selling directly to customers, you can offer your rabbits in a variety of cuts to make the meat more appealing to home cooks.

Rabbit does not have a strong flavor, and its taste is comparable to chicken. Since rabbit does not have a gamey taste, it is very versatile and can be used in a wide variety of recipes since you do not need to cover up a strong flavor. Rabbits have a high meat-to-bone ratio (even more so than poultry), meaning there is more edible meat on the carcass than bones, thus giving your customers more bang for their buck.

What Is Sustainably Sourced Meat?

Sustainably sourced meat comes from animals that were raised in accordance with humane animal welfare standards and environmentally friendly practices.

HOW IS ANIMAL AGRICULTURE DAMAGING OUR ENVIRONMENT?

The United States Department of Agriculture reports that almost 10 billion livestock animals were slaughtered in 2022. This figure increases each year.

Raising this quantity of livestock requires significant land, feed, and water resources. As these resources become more limited, it's worth considering the environmental efficiency of different protein sources. Agricultural land use changes, including clearing forests for crop farming or livestock grazing, can impact biodiversity.

Different livestock species contribute varying amounts of greenhouse gases to the atmosphere. The environmental impact varies greatly depending on the type of animal, production method, and management practices. When considering environmental footprint, smaller animals like rabbits typically require less feed, water, and space per pound of meat produced compared to larger livestock.

WHAT ARE THE MOST SUSTAINABLE MEATS?

In today's day and age, many consumers are concerned about the environment and are looking for environmentally friendly meat sources. Let's look at six of the most popular farmed meat proteins in North America. Let's start with lamb, the least sustainable meat.

#6 LAMB — According to a study performed by the *Meat Eaters Guide to Climate Change and Health*, lamb has the highest carbon footprint. Before being processed and packaged, lamb produces an average of 20.44 kg of CO_2 emissions per kilogram. However, this figure does not take into account the emissions produced as the product is transported from the farm to your plate.

Sheep are ruminant animals, which means that they release methane gas through belching and waste. Methane gas is one of the most harmful greenhouse gases, even more so than carbon dioxide.

#5 BEEF — Beef may produce 5 kg fewer CO_2 emissions per kilogram than lamb, but cattle farming has a massive carbon footprint. Cattle use billions of gallons of water and feed each year, and like sheep, they are ruminant mammals, meaning they produce methane gas as an unwanted by-product.

#4 PORK — Pork is slightly better than beef, but your bacon is not eco-friendly. A whopping 4.62 kg of CO_2 comes from pig farming. However, its carbon footprint greatly increases as meat is transported and processed into bacon, cured ham, or other types of cold cuts.

#3 FARMED SALMON — Salmon farming is one of the most uncontrolled farming industries in the world due to waste, pesticides, and other harmful chemicals leached into the sea. These harmful pollutants destroy habitats, seep into drinking water sources, and kill other types of underwater life. Salmon farming requires huge amounts of energy and produces 4.15 kg of CO_2 per kilogram of salmon.

#2 TURKEY AND CHICKEN — Poultry does not produce methane gas, and it definitely requires less food and water than sheep, pigs, and cows. Chicken and turkey produce 2.33 kg of CO_2 per kilogram of meat before transport and processing. However, there are some serious environmental concerns with the methods used in slaughtering and processing. For example, slaughtering all types of poultry is more energy-intensive than ruminant mammals.

Another undesirable side effect is poultry manure. Poultry manure contains toxic levels of heavy metals, pesticide residues, and harmful bacteria that can damage local soil and waterways. For this reason, turkey and chicken landed second on this list even though they produce less CO_2 than rabbits.

#1 RABBIT — Rabbits are herbivores. A standard rabbit produces an average of 3.6 kg of CO_2 per kilogram of live rabbit. However, studies show the majority of methane emissions produced by rabbits are through their manure. As mentioned before, rabbit manure is extremely beneficial as a natural compost or natural fertilizer.

Another reason why rabbit meat landed number one on the list of most sustainable meat sources is that rabbits convert food and water into edible meat four times more efficiently than sheep and cattle and 1.4 times more efficiently than pigs. Thankfully, these benefits do not come at the expense of flavor.

SUSTAINABLE MEAT SOURCES ARE MORE THAN JUST EMISSIONS

Often, a lack of sustainability becomes synonymous with greenhouse gases, climate change, and carbon emissions. While these are all very important aspects, there is so much more to sustainable farming.

To honestly assess how sustainable and environmentally friendly rabbit farming is, we need to factor in the impact on the land, water, and biodiversity. The way the rabbits are raised and farmed and the impact these methods have on the environment are of equal importance to the carbon footprint itself.

Here are some ways raising livestock can have a negative and positive impact on the environment.

Organic farming

Organic farming focuses on sustainability and biodiversity, revolves around using natural resources, and prohibits the use of synthetic substances. Organic farmers strive to raise their livestock in natural conditions, using wholesome, organic feed. Hormones used to stimulate growth are strictly prohibited, and livestock must be raised in accordance with certain rules to be certified as organic (regulations may vary from state to state).

It is widely accepted that organic farming provides a long list of environmental benefits, but there are some serious concerns as to whether the animals' welfare is better than conventional farming.

Industrial farming

Industrial farming caters to the nation's increasing demand for meat. Often referred to as "factory farming" or "commercial farming," this method focuses on maximizing the output for the least amount of input, thus cutting costs to increase profits.

This type of farming often holds a large number of animals in a cage, pen, or feed lot. The animals are fed in place with restricted mobility, and in most cases, there is almost no access to natural land or light.

Industrial farming is constantly being criticized for its overuse of hormones and antibiotics, genetic manipulation, and lack of regard for the animals' welfare. This method of farming is one of the main reasons for decreased public health, land degradation, increased pollution, deforestation, and poor water management.

Regenerative agriculture

Regenerative agriculture focuses on regenerating soil, biodiversity, and maintaining water systems by conserving and rehabilitating it. Its goal Is to use livestock and manure to improve soil quality. Research shows that regenerating land by rearing livestock in this way actually increases the ability of the ground to absorb excess carbon emissions from the atmosphere.

SUSTAINABLY SOURCED RABBIT MEAT

After having raised a variety of different types of livestock on our homestead, we set up a rabbitry. One aspect that grabbed our attention about producing our own rabbit meat was the sustainability of the entire process. Rabbits are an ecological and sustainable meat source for a variety of reasons.

Efficient

Rabbits are four times more efficient than cattle at converting their feed into protein. While rabbits can consume grass and other forage, they require a balanced diet for proper growth and health. For optimal meat production and consistent growth rates, a high-quality commercial pellet (typically 16-18% protein) should form the foundation of their diet, which can be supplemented with hay and limited amounts of vegetables.

Plus, rabbits are cost-efficient to establish. To get your rabbitry started, you could use secondhand cages. Just be sure that they are secure from predators and are protected from the weather, such as rain, cold, wind, and heat. Other than cages, your bunnies will need water bottles or crocks, feeders, a nest box, and appropriate feed.

They need very little space

Due to their size, rabbits do not require much space. You can place your rabbitry in hutches on the grass, have them stacked in cages, or place them in colonies. Colonies are often used to give rabbits the chance to run, hop, and engage in natural behaviors they love. Additionally, advertising rabbits as free-range instead of caged can be a valuable marketing advantage.

Easy to care for

Rabbits are quiet and clean, which means they will not bother the neighbors if you are raising them in an urban area. I have even heard of people raising rabbits inside their houses. (I do not recommend this method!)

Healthy

Rabbits are hardy mammals and rarely get sick, which means less need for medicine or expensive visits to the vet.

They are reproductive miracles

The idiom "breeding like rabbits" could not be truer! Rabbits have a very short gestation period, between 30 to 35 days, which means they can comfortably have six litters or more a year. An average litter has 7–10 kits, and a mama rabbit can get pregnant almost immediately after giving birth.

Let's say mama rabbit has six litters a year, and each litter has 8 healthy kits, in case some of the babies do not survive. In one year, that would be 48 rabbits per doe per year; that could equal between 140 and 160 pounds of meat each year.

Cruelty-free

Young rabbits, often called "fryers," are typically processed at 8 to 10 weeks of age (per the ARBA Standard of Perfection), as their meat is white and tender. Before you get judgmental about slaughtering young rabbits, these animals are already at or approaching the reproductive stage in their life span. And for comparison, that tender chicken you purchase at the grocery store is normally only five to seven weeks old!.

Rabbits poop a lot!

One thing a rabbit does a lot is poop. Your rabbits might be for eating, but their manure is an amazing compost for your garden, lawn, and flower beds. Rabbit manure equals a larger, healthier yield in your vegetable garden. Plus, many people will pay top dollar for bunny poop.

Rabbits are raised for three main purposes: meat, manure, and fur. The fur trade in North America is highly regulated; for this reason, we will only be discussing the first two in this book.

Raising rabbits for profit can be straightforward, though there are various production methods to consider. While grass can supplement a rabbit's diet, a balanced feed program including quality pellets is essential for optimal growth and development. For meat production, rabbits are typically ready for processing when they reach the appropriate weight—fryers at 8–10 weeks (up to 5.5 pounds) or roasters at 10 weeks to 6 months (up to 9 pounds). Regulations, permits, and inspections vary from state to state, so be sure to process them in accordance with your county or state requirements.

Photo Courtesy of Melina Anderson, Shining Light Farm

When your rabbits are around 12 weeks old, they will be ready for processing. Regulations, permits, and inspections vary from state to state, so be sure to process them in accordance with your county or state.

In North America, rabbits are often considered similar to poultry in meat processing situations due to their size. Play it safe and research what is permitted and what is not to avoid unwanted fines.

There are two ways to process your meat rabbits:

1. By yourself, on your farm — See Chapter 12 (How to Humanely Butcher a Rabbit)
2. At a USDA-inspected facility

Permits and regulations vary among states and counties in the US. For example, in Kentucky, rabbit meat does not need to be USDA-inspected to be sold for human consumption, but the state of Indiana requires rabbit meat to be processed and inspected in an official establishment. In Chapter 12, we will discuss how rabbit meat is inspected.

Where you finally decide to process your rabbits depends entirely on how many rabbits you plan on processing.

Once your rabbits are processed, you should market them to the consumer as soon as possible.

Here is a short list of suggestions for where you can sell your meat:

- Friends and family
- Farmers markets
- High-end restaurants
- Grocery stores
- Online (unfortunately, involves loads of red tape if you are selling across state lines)
- Dog owners who feed their pets a raw food diet
- Commercial pet food companies

On average, rabbit meat sells for $6 per pound and up. Rabbit meat is in high demand for dog food, so you could probably ask for more and get it. Also, if you are marketing your rabbits as grass-fed, you can get more than $10 per pound.

As a side business, you can sell ready-to-use fertilizer made from your rabbit manure. Rabbit manure is a nutrient-packed fertilizer and can be directly applied to plants without the risk of killing the vegetation. So, instead of waiting months to use the manure as fertilizer, as with other animals, rabbit manure can be used right away (and sold!)

Here is a short list of suggestions of who you can sell rabbit manure to:

1. Gardeners
2. Hobby farms
3. Landscaping companies

Rabbit manure can go from $35 to $45 for a four-pound bag. If the manure has been aged or composted with worms, it can go for more. It depends entirely on your location and demand. Many rabbit farmers can make a killing if they know how to market their rabbits' poop correctly.

ASK THE EXPERTS

If someone is thinking of raising rabbits for meat, what considerations do they need to make before they start?

Before starting a rabbit farming journey, there are several key considerations to keep in mind, as shared by our panel of experts. From housing needs and breed selection to understanding the commitment involved, these factors will help determine whether raising rabbits for meat is the right fit for you. The following bullet points summarize the most important aspects to consider—covering the practical setup, financial planning, and emotional readiness that successful rabbit farming demands. This introduction will give you a snapshot of what lies ahead, so you can prepare effectively and ensure a rewarding experience.

1. HOUSING & CLIMATE ADAPTATION

Housing is one of the primary concerns. Rabbits need protection from extreme temperatures—specifically, they thrive in cold but suffer in heat. Many breeders emphasized the need for shaded, draft-free environments, and adaptable setups that can keep rabbits cool in hot climates. Depending on location, housing might need to be insulated to withstand cold winters, with careful attention to ventilation in hot climates to prevent overheating, which can lead to sterility and death.

"Housing is one of the main considerations for keeping rabbits. Whether or not you're keeping them inside or outside, rabbits are sensitive to temperatures, specifically anything above 75°. They tend to thrive in cold weather but anything over 75° they start having issues overheating."

Jeffrey N B Jenson
High Country Farm

2. BREED SELECTION

Breed choice significantly impacts meat production. Experts recommended selecting breeds that grow quickly and have a high meat yield, like New Zealand, Californian, or Rex rabbits. Additionally, temperament and fur quality were also noted as important factors for breed selection.

"Buying good breeding stock is #1. Cheap stock is not good stock. You want genetics for grow outs to be 5lbs by 8 to 10 weeks old. Find out about temperament of the family line. Meanness can sometimes be genetic."

Tabitha Brady
Brady's Bunny's

3. SPACE REQUIREMENTS

Raising rabbits involves ensuring enough space, not only for the breeding stock but also for grow-outs. Many respondents highlighted that beginners often underestimate how quickly space fills up as litters grow. Properly separated cages for breeding, grow-out, and potential isolation (quarantine) are crucial to manage rabbit health and prevent overcrowding.

4. COMMITMENT TO BUTCHERING

Commitment to butchering the rabbits is necessary, as this is an emotional hurdle for many. Several breeders recommended ensuring that everyone in the household is comfortable with the idea. Alternatives like finding a reliable butcher or a nearby farm that can assist with dispatching were also mentioned.

> "It's important to be committed to butchering the animal when it's time or having a reliable service that will do the dispatch for you. If you don't have the determination to plan for this, you will end up with many pets that will continue to multiply."
>
> **Amy Lambrecht**
> Buckeye Rabbits

5. FEEDING, COSTS, & WASTE MANAGEMENT

Feeding rabbits with high-quality pellets, fresh water, and occasional forage is essential. Costs of feeding, setup, and continued care were consistently mentioned as crucial planning aspects. Additionally, waste management is vital—rabbits produce significant waste, which can be used as fertilizer but needs a plan for regular disposal.

6. GOAL SETTING

Understanding the purpose of raising rabbits—whether for self-sufficiency, small-scale commercial production, or as part of a broader farming operation—helps in determining the scale, breed, and setup needed. Several breeders emphasized setting clear goals to match infrastructure and financial investments to meet expectations.

> "Consider what you will do with the non-meat parts. Rabbits have a great fur coat that can be tanned and used to make all sorts of items. Lastly, what breed suits your needs? Obviously you're looking at the meat/commercial breeds but deciding which one best suits your climate."
>
> **Justin Anderton**
> Atlas Homestead

CHAPTER 2

Rabbits 101 — Start Here

Welcome to the basics, folks! In this chapter, we're diving into everything you need to know about rabbits—their quirks, behaviors, and all the little details that make them such fascinating creatures. Before you can successfully raise rabbits for meat, you need to truly understand them—where they come from, how they live, and what makes them tick. Think of it like learning the language of rabbits. The more you understand their natural instincts and needs, the better you'll be at giving them a happy, healthy life, which in turn means better results for you as a farmer.

We'll cover everything from where rabbits originally lived, to their curious reproductive habits, their social behaviors, and why they're smarter than most people give them credit for. These are the fundamentals that will help you become a confident and compassionate rabbit farmer. By the end of this chapter, you'll not only know what rabbits need, but also why those things are important. And that understanding is the key to raising strong, healthy rabbits that thrive in your care. Let's hop right in!

Natural Habitat for Rabbits

Often, rabbits are assumed to be a type of rodent due to their long teeth and affinity for gnawing, but actually, they are part of the Lagomorpha family, which includes rabbits, hares, and pikas, small mountain-dwelling mammals found in Africa and North America.

Rabbits, like other animals from the Lagomorpha family, can be found in grasslands, wetlands, deserts, tundra, meadows, farmlands, and forests. Originally, rabbits were only found in Europe and Asia. But now, wild rabbits can be found on every continent except for Antarctica.

Wild rabbits often make their homes by burrowing into the ground; these tunnels are called warrens and can include different rooms for sleeping or nesting. A typical warren may be almost 10 feet deep and have multiple entries for quick escape. Domestic rabbits are often unable to burrow into the ground to protect themselves from heat exhaustion or hypothermia. So, you will need to ensure that their housing is well protected from the elements.

Characteristics

Most people can easily recognize a rabbit when they see one. Rabbits can be shy, silly, friendly, curious, and high spirited, regardless of their sex or breed. Each rabbit breed is different from the next, but here are some common characteristics that make them so interesting.

RABBIT REPRODUCTION

Rabbits are known for their insatiable reproductive habits. Rabbits in the wild tend to breed at least four times a year. Maybe this is because only 15% of baby bunnies make it to their first birthday, so instinctively, rabbits make more babies to ensure their population grows.

In the wild each female rabbit or doe produces between one to 14 babies, called kittens or kits. By three weeks old, a kit can fend for itself. In two to three months, the rabbit is ready to start a family of its own. If there are no natural predators in the rabbit's area, it can quickly become overrun with bunnies.

INTELLIGENCE

Rabbits are intelligent animals with their own unique cognitive abilities. While their intelligence manifests differently than that of cats or dogs, rabbits can learn, remember, and develop complex social behaviors. Rabbits have distinct personalities and may resist handling or activities that make them uncomfortable, which requires patient and consistent interaction.

Some breeds can be trained to recognize their names and basic commands, such as coming when called. Rabbits have an excellent memory and do not easily forget good nor negative experiences. Your rabbits, even though they are raised for meat, will come to recognize your voice, scent, and even your facial features. Due to rabbits' curious personalities, if you use a gentle and soft voice with them, they will often come over and inspect you.

However, their intelligence means they benefit from appropriate environmental enrichment. While rabbits adapt well to properly designed cage systems, they also benefit from interaction and enrichment. For production settings, many rabbit farmers provide proper cage space with occasional enrichment items like gnawing blocks. Some operations utilize colony or pen systems that allow for more natural behaviors, though these require careful management to prevent fighting and disease transmission. Regular, calm handling and interaction helps maintain docile animals that are easier to manage.

APPEARANCE

Rabbits have a stout body, a round back, long ears, and large back legs for hopping around. They also have a cute fluffy tail and long front teeth. Here are some interesting facts about your rabbits' appearance.

Ears

Among meat breeds, a rabbit's ears vary in length depending on the breed, with many exceeding four inches. Research has discovered that the ear length not only helps them to hear predators and friends but is crucial for thermoregulation. The large surface lets bunnies release excess heat and stay cool, which is why rabbits that live in the desert have the longest ears. Commercial meat breeds typically have ear lengths proportional to their body size, as specified in the Standard of Perfection.

It turns out rabbit ears can detect a wide range of frequencies. Their hearing is so sensitive that they can hear ultrasonic sounds from bats. Rabbits can clearly hear frequencies between 96 Hz and 49,000 Hz from up to 1.8 miles away. While rabbits are sensitive to sounds, they can adapt to normal environmental noise. It's best to avoid placing their housing directly next to extremely loud equipment like radiators or generators, but rabbits can acclimate to typical farm and household sounds. In fact, exposure to normal activity sounds helps prevent rabbits from becoming startled by routine noises.

DID YOU KNOW? Flop-eared rabbits are actually hard of hearing or even deaf. The anatomy of the flop means the ear canal has a kink, causing the canal to narrow and preventing sound from traveling down to the eardrum easily.

A rabbit on high alert will have its ears sticking straight up and listening. If you watch carefully, you will notice that they rotate their ears 180 degrees to pinpoint the exact spot from which the sound is coming.

Eyes

Rabbits' eyes, which can rotate almost 360 degrees, are located high and near the top of the skull. In addition, rabbits are farsighted, equipping them with the ability to spot predators in any direction. Due to being crepuscular creatures (meaning they are most active at dawn and dusk), rabbits can see well in low-light conditions. This helps them forage for fresh weeds and grass. For this reason, many rabbit farmers find their animals are most active and ready to eat in the evening and early morning.

Eye color varies widely among meat rabbit breeds. While some breeds like New Zealand Whites and Californians have red eyes due to their genetics, many meat breeds come in various colors with dark eyes. Red-eyed (albino) rabbits may have somewhat different visual abilities—they typically have poorer depth perception and

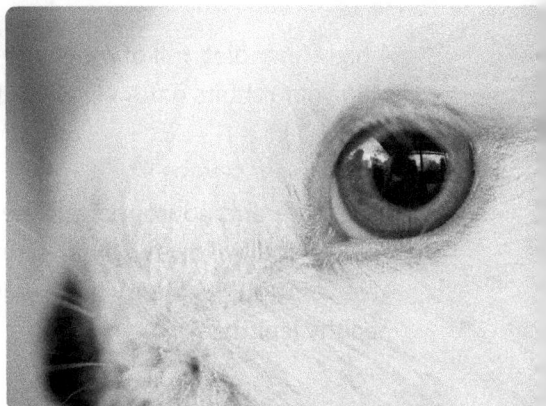

may be more sensitive to bright light due to the lack of pigmentation in their irises. When raising red-eyed varieties, providing some shelter from direct, intense sunlight can be beneficial, though they adapt well to normal lighting conditions in proper housing.

DID YOU KNOW?
Rabbits have transparent eyelids that they use while they sleep, which act as a defense mechanism. If a predator approaches, the light receptors in the eye will send a signal to the brain about the danger, causing the rabbit to snap into motion quickly.

Teeth

Rabbits are known for their big teeth. They will nibble and gnaw at basically anything they can get their teeth on. But why do their teeth always seem to be growing?

Rabbits' teeth seem to be constantly growing because they are! They need sharp teeth to deal with their high-fiber diet. Rabbits' teeth are designed to never stop growing. Generally, their teeth grow around one centimeter per month. This isn't an issue for bunnies being used for processing, but your older rabbits used for breeding purposes will need their teeth ground down. There are some things you can do to help them do this:

High-fiber diet

A high-fiber diet, full of roughage, such as Tomothy or Orchard hay, will help your rabbits naturally wear down their teeth.

Encourage chewing

Offer safe and appropriate chew toys or branches, such as untreated apple wood, willow twigs, pinecones, or pieces of pine sticks. These will give your rabbits plenty of gnawing pleasure and keep their teeth at a healthy length.

Include leafy greens

Provide an ample supply of leafy greens, such as kale, parsley, cilantro, and dandelion greens. Chewing on fresh greens stimulates the jaw and promotes dental wear.

Limit processed treats

Avoid excessive sugary or starchy treats, as they contribute to dental issues. If you want to give your breeders a special treat, opt for healthy snacks, such as small pieces of fruit or vegetables.

It is important to note that some rabbits are prone to dental issues, such as malocclusion. Malocclusion is when a rabbit's teeth become misaligned. While a proper diet with sufficient roughage helps maintain dental health, malocclusion is primarily caused by genetic factors. The inherited jaw structure determines how the teeth align and grow. When selecting breeding stock, examining dental alignment is important to reduce this trait in your herd.

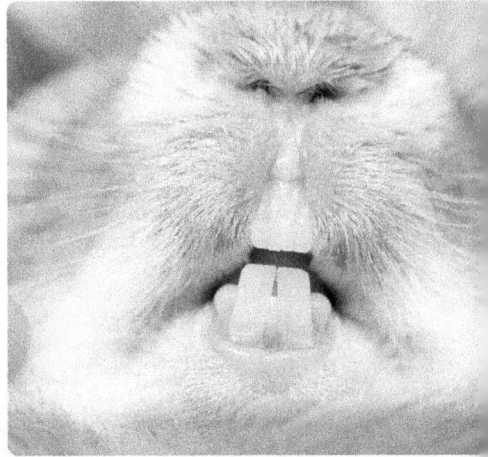

Malocclusion can be fatal, as eventually, the rabbit will not be able to eat. If a rabbit cannot eat, its digestion system will stop moving, causing a condition called gut stasis. Gut stasis is fatal in rabbits. Here are some signs of malocclusion in rabbits.

- Teeth growing at odd angles, such as into or out of the mouth
- Sores or abscesses inside the mouth
- Drooling
- Unexplained weight loss
- Less poop being produced
- Pawing at mouth
- Difficulty eating

If you notice that your mature breeding rabbits have misaligned teeth, they should be processed and packaged as fryers to avoid the chance of passing on this defect to future rabbits.

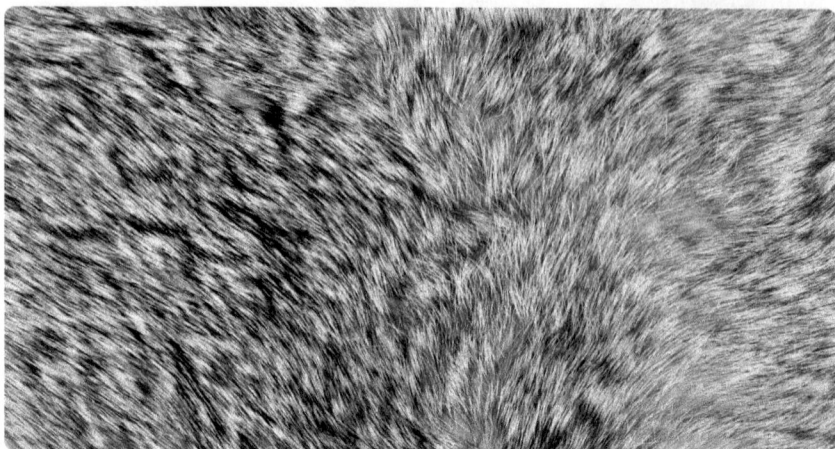

Coat

Domesticated rabbits come in a variety of coat colors; wild bunnies are typically brown or tan. This coloration helps rabbits to blend into their natural habitat to avoid being easily spotted by predators.

Some commercial meat rabbits like the Florida White and Californian have white fur, which is often preferred in certain segments of the meat industry. White-furred rabbits are favored in some commercial meat production settings for several reasons.

1. White fur is easier to detect during the processing and packaging stages. This improves the visual appeal and cleanliness of the final product.
2. White fur is less likely to leave pigment marks on the meat, which can occur with darker-colored fur.
3. White fur provides a consistent and uniform appearance in the colony. This makes it easier for producers to assess the health and quality of rabbits at a glance.
4. White fur reflects more sunlight and heat compared to darker fur. This can help keep the rabbits stay cooler in hot climates or during the summer months. Managing your rabbits' temperature is important for their well-being and growth.
5. With a large group of identical white rabbits, it may be easier for you to remain detached and avoid picking a favorite rabbit, which can make the butchering process less emotionally challenging.

It is worth noting that while white-furred rabbits are predominant in commercial meat production, there are still many meat rabbit breeds with different coat coverings. Ultimately, the choice of rabbit breed and fur color may vary depending on regional preferences, market demands, and specific production goals.

BEHAVIOR

Rabbits are social creatures and in the wild live in groups called colonies. Rabbits are crepuscular, meaning they are most active at dawn and dusk, as these times allow them to forage while minimizing exposure to predators. Common predators for rabbits are owls, hawks, eagles, feral cats, falcons, dogs, foxes, coyotes, and bobcats. Thankfully, rabbits have long legs that let them run at high speeds for extended periods of time to help them escape.

Rabbits are cunning. To outrun a predator, a cottontail rabbit will run in a zigzag pattern and can reach up to 18 mph. If one of your domesticated rabbits happens to escape, be sure you have your sneakers on and maybe some helping hands to catch it.

> *"I tell everyone to keep a good fishing net where you can reach it for catching escapees!"*
>
> **Rachel Heaton**, Hardly Simple Farming

In the wild, rabbits spend most of their time digging, hopping, running, eating, stretching, and socializing with other rabbits. To ensure healthy rabbits, be sure that they can carry out these natural behaviors in their housing units.

Rabbits have quite a few quirky behaviors, and here is a little insight to help you decipher them.

Why do rabbits rub their chins on things?

This behavior is called "chinning" and is quite normal. Rabbits, like dogs and cats, have scent glands on either side of their anal openings, in their eyes, inside of their mouths, and under their chins. Each gland has different functions; for example, the chin and anal glands produce a waxy substance that has chemical messengers called pheromones. These pheromones deliver messages to let other rabbits know they are ready to mate or claim ownership.

Why does my rabbit make strange grunting noises?

Rabbits grunt when they feel threatened or angry. Sometimes, the grunting is followed by thumping or stamping the hind legs or even a bite or nip. Often, rabbits will be upset while you rearrange their cage as you clean; they may grunt, nip, or even charge at you. Rabbits are creatures of habit and do not like change. They will calm down once everything is put back where it should go.

How do I know if my rabbit is happy?

A happy rabbit will often do a bunny dance. This involves leaping, hopping, jumping, spinning in the air, and even racing around. Some rabbit farmers try to give their rabbits 30 minutes a day of playtime in a pen to allow them to stretch and burn off their excess energy.

What does it mean when my rabbit nips or bites me?

Bunnies nip to get your attention or to ask you to move out of their way. Typically, a bunny nip will not hurt. On the other hand, a rabbit bite is a little more painful. There are a few reasons why your rabbit may bite you. For example, rabbits have poor up-close vision, and maybe you surprised them, and they mistook your hand for a predator or something yummy to eat. If your rabbit bites you, let out a shrill cry. Rabbits tend to cry out when hurt and will be surprised by your behavior and normally stop biting after a few times. Any rabbit that continues to bite should be culled from your herd, and some breeders have a zero-tolerance policy with biters.

Why is my rabbit spraying?

Both male and female rabbits may spray, as it is a way to mark their territory. This is not common with meat rabbits, as they are processed before they reach sexual maturity, but rabbits used in breeding often spray, though not all will exhibit this behavior. Male rabbits may mark their territory and their females by urinating on them. Female rabbits sometimes spray their territory to signal to males that they are receptive to breeding. While neutering and spaying generally reduces spraying behavior, a small percentage of altered rabbits may still occasionally spray.

Why is my rabbit eating its droppings?

Rabbits are herbivores. Herbivores often regurgitate partially digested food (cud) and chew on it to break it down. But rabbits have a different method called hindgut fermentation.

How does it work? Once the rabbit eats its food, the food goes through a fermentation process in the digestive tract called the cecum, which produces special fecal matter called cecotropes or "night feces." This fecal matter is often only produced once a day and is produced like regular fecal matter but softer. These feces are rich in nutrients and protein and have a higher concentration of vitamins than the typical hard bunny droppings you see in your rabbits' cage. Cecotrope ingestion is normal and an important part of your rabbits' behavior.

LIFE SPAN

The average life span of domesticated rabbits can vary depending on various factors, including breed, genetics, diet, health care, and living conditions. On average, domesticated rabbits live between eight to 12 years. However, rabbits for breeding purposes, such as those raised for commercial meat production, typically have a shorter life span compared to companion rabbits. On average, breeding rabbits may live between five and eight years, but this can vary greatly based on individual circumstances.

Breeding rabbits have a shorter life span due to the fact they may experience more physical stress from reproductive demands. Rabbit farmers will need to provide their breeding rabbits with high-quality nutrition and follow appropriate breeding management to support the health of the rabbits and their litters.

Caring for Your Meat Rabbits

Raising meat rabbits involves more than processing them for sale; it is important for you to understand how to properly care for them. Remember, rabbits being raised for meat need and deserve a happy life. Not only does it provide better nutrition for your customers, but it is the right thing to do. Here are some suggestions for giving your meat bunnies the best possible life, even if it's a short one.

Clean their house

No matter what style of housing you choose to use for your meat rabbits, their cages will need cleaning. From time to time, excess hay will cause a blockage, preventing the rabbits' waste from falling through the wire. Most likely, you will need to use a hand-held garden hoe to break up and remove any old hay. In Chapter 6, we will go into more detail on how to clean your rabbits' housing.

Photo Courtesy of Toni Case, Lost Mountain Farm

Protect those bunny ears

Rabbits have sensitive ears, and hay has ear mites. Ear mites are tiny bugs that often live in hay and sometimes find a new home in a rabbit's ear canal. It causes a lot of wax and scabs. It looks painful, but it is really extremely itchy. Treatment is ear drops for 10 days. To prevent a full-blown infection, I put a couple of drops of baby oil in my rabbits' ears once a week. If that is too much work, you can use mineral oil or vegetable oil (something you don't mind the rabbit ingesting), and give each ear about a tablespoon worth, once. That results in a very greasy rabbit for about two weeks, but there is usually no need to repeat treatment.

Protect their feet

The comfort of rabbits in wire-floored cages depends on several factors. Most commercial breeds have dense fur pads on their feet that protect them when housed on the correct gauge of wire (typically 14×1" for adults). However, some breeds, particularly Rex varieties with different fur structure, may be more prone to developing sore hocks. For these breeds or for heavy rabbits, providing a resting board or mat gives them an alternative surface. Regular cage cleaning, proper wire gauge selection, and keeping toenails trimmed are important preventative measures for maintaining foot health in wire-floored housing systems.

Nesting places

There is no need to leave nesting boxes in cages all the time. They take up a lot of space that your rabbits would rather have to move about. I only put a nest box in the cage if I know the doe is ready to give birth. If I do place a nest box in the cage, I put fresh hay inside and promptly remove any spoiled hay. It is like a baby sitting in a soiled diaper; you do not want rabbits sitting in their own waste.

Clean water

Rabbits need access to an ample source of clean, fresh water. Rabbits drink lots of water, and even more when they are pregnant and nursing. Be sure to check your rabbits' waterers daily in case they get clogged. Some rabbits cannot use a rabbit nozzle on their waterers and may need to use a water bowl, so use what works for your rabbits individually.

Best nutrition

Make sure your rabbits are provided a wholesome diet. I tried feeding my rabbits only grass and hay, but it took several weeks longer for them to reach an adequate weight to be processed. Now, I feed my rabbits both pellets and hay daily. Pellets are an excellent source of protein, and I add about a handful of hay—about the size of a rabbit's head. Plus, every night, I give them a small handful of fresh greens, such as kale, parsley, grass, etc.

Exercise

Nobody wants fat rabbits. Providing your bunnies with exercise can be accomplished in a number of different ways. You can use a puppy playpen in a place where your rabbits are safe from predators. Or you can use a rabbit tractor and let them hop about while they munch on grass. I feel my rabbits prefer hutches because it gives them a sense of security. But giving them time outside their hutches is important for their overall health.

Watch for fur mites

Fur mites are tiny parasites that can cause itching, hair loss, and skin irritation in rabbits. To keep them at bay, regularly inspect your rabbits' fur for signs of mites, especially around the neck and back. If you spot any signs of infestation, treat promptly with a vet-recommended mite treatment to prevent discomfort and the spread to other rabbits. Keeping housing clean and providing good hygiene practices helps reduce the risk of fur mites.

My meat rabbits are not pets and are not treated as such; however, I do treat them like well-cared-for livestock. Personally, I need a degree of separation because my meat bunnies are a source of income for my family.

Please be sure to take care of your rabbits and whatever needs they may have.

Common Questions Asked by First-Time Rabbit Raisers

Commercial rabbit farming is not for the faint of heart, as they are living animals and require time, specialized housing, diet, and health care. If you are serious about rabbit farming, you will need to invest your time, energy, and money into your operation to ensure your bunnies' health and welfare.

What are some challenges to rabbit farming?

Raising rabbits is pretty straightforward when compared to other types of livestock, such as poultry and cattle. However, all living creatures are susceptible to disease. Diseases can be easily spread through contact with other symptomatic rabbits, contaminated food or water, or even an infected human. Another problem is the weather. Extreme weather conditions can make your rabbits very sick or may even be fatal. Heat waves, deep freezes, and even tornadoes and hurricanes can be deadly for rabbits.

How much space is required for my rabbitry?

Space is essential for a successful rabbitry. The amount of space required depends entirely on the type of housing system you will be using and the number of rabbits you plan to raise. If you plan to raise your rabbits outside in pens instead of hutches, logically, you will need more space to let your rabbits graze on pasture. A pasture plat for one rabbit should be at least 50 sq. ft, as this will provide the rabbits plenty of room to forage and hop about freely.

How difficult is the upkeep of a rabbitry?

Maintaining a rabbitry is not as difficult as you might imagine. First of all, you need to choose a well-ventilated area with sunlight. If you are starting off small, make sure the space is large enough for the number of rabbits you plan on raising and not just the starting number of rabbits. Later on in this book, we will discuss some simple techniques to keep your rabbitry clean.

One of the best ways to quickly increase your rabbits' population is to provide them with adequate space to live and breed. Another key element is to provide your bunnies with a wholesome diet rich in nutrients that will help them produce healthy litters. Finally, make your mama rabbits comfortable by giving them ample space and materials to build cozy nests for their babies. Happy rabbits produce more happy rabbits!

There are quite a few factors that affect your rabbits' growth; however, the main factor is diet. Rabbits, like humans, require a diet rich in nutrients in order to grow and develop properly. Not surprisingly, the second factor is exercise. Exercise is essential for keeping your rabbits' bones and muscles strong. Plus, you must make sure your rabbits always have access to a clean water supply to prevent them from becoming dehydrated.

This is a topic among rabbit farmers—to vaccinate or not. In North America, there are no required vaccines for rabbits regardless of their age or purpose. Vaccinations for rabbits are more common in Europe, where certain diseases like Rabbit Hemorrhagic Disease Virus (RHDV) are prevalent. Many breeders focus on selecting for naturally healthy stock and implementing good biosecurity practices rather than relying on vaccines. This approach helps develop rabbits with stronger natural resistance to common ailments like Pasteurella.

Most meat rabbit breeds reach fertility between four to six months of age, with optimal breeding age varying by breed and individual development. Generally, does can be bred when they reach 75-80% of their adult weight, which typically occurs between 5-7 months for most commercial breeds. Waiting too long (beyond 7-8 months) can sometimes lead to reduced fertility as does may develop fat deposits around their ovaries. The timing should balance physical maturity with reproductive efficiency, while also allowing you to evaluate their health, temperament, and conformation traits.

What should I do after the doe has given birth?

Monitor the nest closely. Keep the nest clean and free of debris, as this will prevent the spread of disease or infection. Promptly remove any stillborn bunnies. Restrict all animals and visitors from entering or even roaming about the rabbitry as this will cause undue stress to new mothers and kits.

Photo Courtesy of Angel Wills, Happy Hill Farmstead

CHAPTER 3

Choosing Your Breed

A ll rabbit breeds have strong points and weaknesses. Purchasing the best breeding stock possible will help to ensure your rabbitry's economic success. But you also need to make sure the breed will fit your needs for meat production. In this chapter, we will discuss the pros and cons of some of the most popular purebred rabbit breeds for meat production and how to choose the correct breed for your rabbitry.

Different Rabbit Breeds

The technical term for breeding and raising domesticated rabbits is cuniculture. Cuniculus is the Latin word for rabbit. Humans have been practicing cuniculture for centuries. There are more than 300 different rabbit breeds worldwide; however, ARBA only recognizes 52, which can be divided into three different categories for cuniculture.

Angora Rabbits

RABBITS FOR WOOL

Often, when we think of wool, we instantly think of sheep; however, fluffy wool can come from many different types of animals, such as goats, llamas, alpacas, and even rabbits. Angoras are the most common type of rabbits used for wool production due to their soft, fluffy wool.

RABBITS FOR FUR AND PELTS

In the past, rabbits were specifically bred for the purpose of fur trading. Typically, fur rabbits have short, shiny coats instead of long fluffy wool. Rabbit pelts with the fur still attached are sold to be made into a variety of clothing items, such as hats, scarves, gloves, and so on. Rex, satin, and silver fox rabbits are three of the most common breeds used in fur production due to their desirable fur qualities.

Rabbits for Meat

Finally, it is time to talk about which rabbit breeds will give you the biggest bang for your buck. Choosing the best breed means knowing what to look for, and unfortunately, those adorable little lionhead rabbits are not your best choice. They are basically the equivalent of the size of a Cornish hen compared to a chicken.

There are three factors to take into consideration when choosing the best rabbit breed for meat:

1. Growth rate
2. The doe's mothering skills
3. Meat-to-bone ratio

GROWTH RATE

To minimize expenses, you should look for breeds that grow quickly and efficiently. The longer you have to care for your rabbits before processing, the more money and time out of your pocket. For commercial meat production, rabbits are typically processed according to specific weight and age categories: fryers (up to 5.5 pounds, under 10 weeks), roasters (up to 9 pounds, between 10 weeks and 6 months), and stewers (over 6 months). The most economical production usually focuses on fryers, which provide the most tender meat and best feed conversion. The ideal processing time and weight will depend on your market and production goals.

THE DOE'S MOTHERING SKILLS

The doe must have exceptional mothering skills to routinely care for litters of eight or more kits. Good mothers will give you healthy grow outs for months. Unfortunately, not all meat rabbits exhibit good skills in caring for their young, which can end up costing you in the long run.

MEAT-TO-BONE RATIO

In the United States, a fully dressed rabbit should weigh between three to four pounds. Many breeders cross breeds to reach a higher bone ratio of four to five pounds; for example, they cross New Zealand Whites with Californians or Flemish Giants with New Zealand Whites to produce larger, more efficient meat rabbits.

The following list consists of the most common rabbit breeds raised for meat.

Californian

The Californian rabbit is also known as the Californian White. This breed was developed in the early 1920s by George West when he crossed purebred New Zealand Whites with Chinchilla and Himalayan rabbits. The Californian rabbit is one of the world's most popular rabbit breeds for commercial production.

MEAT PRODUCTION
Californian rabbits are primarily raised for meat production. They have a good meat-to-bone ratio, meaning they produce a good amount of meat relative to their bone structure. Their meat is described as tender, juicy, and mild in flavor.

GROWTH RATE
Californian rabbits have a relatively fast growth rate compared to some other rabbit breeds. They can reach market weight within a reasonable time frame, making them efficient for meat production.

FEED CONVERSION EFFICIENCY
Californian rabbits are known for their efficient feed conversion, meaning they can convert feed into body weight effectively. This can result in cost savings on feed expenses.

DOCILE TEMPERAMENT
Californian rabbits generally have a calm and docile temperament which can make them easier to handle and manage. This is beneficial for both breeders and those handling the rabbits. This hearty breed is also known for creating great mothers.

HEAT SENSITIVITY
Californian rabbits may be more sensitive to extreme heat due to their dense fur and limited ability to dissipate heat They may require additional care and measures to keep them cool and comfortable during hot weather.

SPACE REQUIREMENTS
Though Californian rabbits are a medium-sized breed, they still require adequate space for proper housing and exercise. Providing sufficient space for them to move around is important for their well-being.

AVAILABILITY
While Californian rabbits are a recognized and popular breed, their availability may vary depending on the location and demand. Finding reputable breeders and acquiring breeding stock may require some research and effort.

New Zealand White

This breed is one of the world's most popular breeds for meat production. New Zealand White rabbits have red eyes, pink noses, and white pelts due to their genetic makeup. While the white coloration is often associated with albinism, it's simply one of several color varieties in the New Zealand breed. New Zealands are also recognized in Red, Black, Blue, and Broken varieties, all of which retain the breed's excellent meat production qualities.

+   **✕**

MEAT PRODUCTION

New Zealand rabbits are primarily raised for meat production. They have a good meat-to-bone ratio, meaning they produce a substantial amount of quality meat relative to their bone structure. The meat is appreciated for being lean, succulent, and flavorful.

GROWTH RATE

New Zealand rabbits are known for their fast growth rate. They have efficient weight gain, allowing them to reach market weight relatively quickly compared to some other breeds.

FEED CONVERSION EFFICIENCY

New Zealand rabbits are efficient at converting feed into body weight. This makes them cost-effective to raise as they can make the most out of the feed provided.

ADAPTABILITY

New Zealand rabbits are adaptable to various climates and environments. They can tolerate different conditions, making them suitable for a wide range of locations.

SPACE REQUIREMENTS:

New Zealand rabbits are a medium-sized breed, and they still require adequate space for proper housing and exercise. Providing sufficient space for them to move around and stretch their legs is important for their well-being.

HIGH FEED INTAKE:

Due to their rapid growth rate, New Zealand rabbits have a relatively high feed intake compared to smaller breeds. This can result in higher feed costs, especially if you are raising them on a large scale.

HEAT SENSITIVITY

New Zealand rabbits, like most rabbits, may be more sensitive to extreme heat. They may require additional measures to keep them cool and comfortable during hot weather, such as providing shade or proper ventilation in their housing.

Rex

Rex rabbits are often sought for their unique fur texture. They are also highly regarded for the excellent quality of their meat due to its fine-textured flesh and outstanding taste.

MEAT QUALITY

While the primary focus of Rex rabbits may not be meat production, their meat is often praised for its tenderness and flavor.

DOCILE TEMPERAMENT

Rex rabbits are generally known for their calm and friendly temperament.

LOWER GROWTH RATE

Compared to some commercial meat rabbit breeds, Rex rabbits may have a slower growth rate. They may take longer to reach market weight, which can affect their suitability for large-scale meat production.

HEAT SENSITIVITY

Rex rabbits may be more sensitive to extreme heat due to their dense fur coat. Extra care should be taken to provide them with appropriate cooling measures during hot weather conditions.

GENETIC HEALTH CONCERNS

As with any breed, there can be specific genetic health concerns that need to be addressed in the breeding and management of Rex rabbits. Working with reputable breeders who prioritize genetic health and maintain proper breeding practices is important.

TAMUK

The TAMUK rabbit is a cross-breed developed by Texas A&M University-Kingsville to withstand hot climates and thrive as a meat rabbit. Created from a blend of New Zealand, Californian, and Altex breeds, TAMUK rabbits were selectively bred for heat tolerance, disease resistance, and efficient growth. This combination makes them an ideal choice for meat production, particularly in warmer regions where other breeds may struggle.

Photo Courtesy of Sierra Hamilton, Crooked Hill Rabbitry

HEAT TOLERANCE

TAMUK rabbits were specifically bred to thrive in hot climates, so they are well-suited for regions with high temperatures. This unique trait helps reduce the risks associated with heat stress, which can be a significant challenge for other breeds.

EFFICIENT GROWTH RATE

With a strong growth rate, TAMUK rabbits can reach market weight in a relatively short time, making them efficient for meat production.

DISEASE RESISTANCE

TAMUK rabbits are selectively bred for resistance to common health issues, including respiratory and digestive illnesses, making them a hardy choice for breeders looking to minimize health risks.

ADAPTABILITY

TAMUK rabbits adapt well to various environments and housing setups, whether they're raised in hutches, colonies, or open-air enclosures. This flexibility is a great advantage for breeders in diverse climates and conditions.

HIGH FEED CONVERSION

Known for their excellent feed-to-meat conversion ratio, TAMUK rabbits can produce quality meat efficiently, making them cost-effective to raise.

AVAILABILITY

As a relatively new breed, TAMUK rabbits may be harder to source outside of Texas and the southern U.S., potentially requiring breeders to travel or order from specialized suppliers.

MODERATE SIZE

While efficient, TAMUK rabbits are medium-sized and may not yield as much meat per rabbit as larger commercial breeds like New Zealand Whites or Californians.

SPACE REQUIREMENTS FOR HEAT

Although they are heat-tolerant, TAMUK rabbits still need well-ventilated housing and shaded areas to maximize their comfort and growth in extreme temperatures.

HIGHER INITIAL COST

Due to their unique breeding and high demand, TAMUK rabbits may come at a higher initial purchase cost compared to more common meat breeds.

Satin

This breed originated in Indiana in the 1930s and has a unique genetic mutation that causes a hollow hair shaft, giving the coat a satin shine, hence its name. The rabbits are valued for their good growth rates, meat-to-bone ratio, and commercial body type. Like other well-established commercial breeds, Satins have benefited from decades of selective breeding for production qualities.

+

MEAT QUALITY

While satin rabbits are primarily known for their fur, they are also raised for meat production. Their meat is often praised for its tenderness and flavor.

EFFICIENT FEED CONVERSION

Satin rabbits are known for their efficient feed conversion, meaning they can convert feed into body weight effectively. This can result in cost savings on feed expenses.

DOCILE TEMPERAMENT

Satin rabbits are generally known for their calm and friendly temperament. They are often considered gentle and easy to handle, making them suitable as pets or for show purposes.

✕

FUR MAINTENANCE

Satin rabbit fur requires more attention and maintenance compared to some other breeds. Their silky fur can be prone to matting and tangling, and regular grooming is necessary to keep their fur in good condition.

HEAT SENSITIVITY

Satin rabbits may be more sensitive to extreme heat due to their dense fur coat. Extra care should be taken to provide them with appropriate cooling measures during hot weather conditions.

Flemish Giant

According to Guinness World Records, the largest rabbit ever recorded was a Flemish Giant named Darius, who was four feet, three inches in length. They are very popular as pets but not as popular for commercial meat production.

+

SIZE

Flemish Giants are known for their large size, making them one of the largest rabbit breeds. This results in a higher meat yield per individual rabbit, making them suitable for meat production.

MEAT QUALITY

The meat of Flemish Giants is often praised for its flavor and tenderness.

DOCILE TEMPERAMENT

Flemish Giants are generally known for their calm and gentle disposition. Their docile temperament makes them easier to handle.

EXCELLENT MOTHERING SKILLS

Flemish Giant does are known for their excellent maternal instincts and mothering skills. They often exhibit good nurturing behaviors, which can be beneficial for raising their young.

✕

SPACE REQUIREMENTS

Due to their large size, Flemish Giants require more space compared to smaller rabbit breeds. Adequate housing and exercise space should be provided to ensure their well-being.

FEED CONSUMPTION

Flemish Giants have higher feed intake requirements compared to smaller breeds. They may consume more feed to maintain their size and growth, which can result in higher feed costs.

SLOWER GROWTH RATE

Compared to some commercial meat rabbit breeds, Flemish Giants have a slower growth rate. They usually take about a month longer to reach market weight, requiring more time and resources for meat production. By the time they meet the desired meat-to-bone ratio, they are considered to be fryers and are harder to market in the commercial market in North America.

HEALTH CONCERNS

Their large size can make Flemish Giants more prone to certain health issues, such as skeletal problems or obesity. Proper care, including a balanced diet and regular health checks, is important to maintain their well-being.

Heritage breeds

Heritage rabbit breeds are defined by the Livestock Conservancy as breeds that are either critical, threatened, watched, recovering, or being studied. These breeds once thrived in North America as a sustainable meat source but have become less and less common during the age of farm industrialization. Heritage breeds are an excellent option for a small, sustainable farm or homestead.

Using heritage rabbits for meat production has its advantages and disadvantages. Here are some points to consider.

FLAVOR AND QUALITY

Heritage rabbit breeds are known for their delicious and tender meat. They often have a richer flavor compared to commercial breeds.

GENETIC DIVERSITY

Heritage breeds contribute to maintaining genetic diversity within rabbit populations, preserving unique traits and adaptations.

HOMESTEADING AND SELF-SUFFICIENCY

Raising heritage rabbits for meat can be a valuable aspect of homesteading and self-sufficiency, allowing individuals or families to have control over their food source.

MARKET DEMAND

Commercial rabbit meat production is dominated by larger-scale operations that often use specific commercial breeds. The demand for heritage rabbit meat may be limited and localized, making it challenging to find a market.

GROWTH RATE

Heritage rabbit breeds may have a slower growth rate compared to commercial breeds. This can result in longer production cycles and higher costs associated with feed and care.

BREEDING CHALLENGES

Maintaining heritage rabbit breeds requires careful breeding management to avoid inbreeding and maintain genetic diversity.

LIMITED AVAILABILITY

Compared to commercial breeds, heritage rabbit breeds may be less readily available. Finding breeding stock and establishing a sustainable breeding program can be more challenging.

LESS UNIFORMITY

Heritage rabbit breeds often exhibit greater variability in size, color, and other characteristics, which may pose challenges in marketing and consistent product presentation.

Be sure to carefully consider these factors when deciding whether to use heritage or commercial rabbits for meat production, taking into consideration your specific goals, resources, and market conditions. Below is a quick overview of some of the best heritage rabbits for meat production.

American

ARBA recognized the American rabbit in 1918 when it was still called the German Blue Vienna. During the early 20th century, the American rabbit was one of the most popular breeds throughout North America, but nowadays, it is considered to be one of the rarest.

+

ADAPTABILITY
American rabbits are known for their adaptability to various climates and environments. They can thrive in different conditions, making them suitable for different regions.

EFFICIENT FEED CONVERSION
They have a reputation for efficient feed conversion, meaning they can convert feed into body weight effectively. This can result in cost savings on feed expenses.

DOCILE TEMPERAMENT
American rabbits are generally known for having a calm and docile temperament, which can make them easier to handle and manage.

X

SLOW GROWTH RATE
Compared to some commercial meat rabbit breeds, American rabbits tend to have a slower growth rate. This means they may take longer to reach market weight compared to other breeds.

MEDIUM SIZE
Although American rabbits are medium-sized, their meat yield per rabbit may be lower than that of larger meat breeds.

LIMITED AVAILABILITY
While American rabbits are recognized as a breed, they may not be as widely available as some other commercial meat rabbit breeds. Finding breeders and acquiring breeding stock may require some effort.

Champagne d'Argent

The Champagne d'Argent is one of the oldest purebred rabbit breeds in the world. It is believed to have originated in the early 17th century in the Champagne region of France. The global population is estimated to be less than 1000.

+

MEAT QUALITY

Champagne d'Argent rabbits are known for their excellent meat quality. The meat is tender, flavorful, and often compared to high-quality poultry.

GROWTH RATE

Champagne d'Argent rabbits have a relatively fast growth rate, allowing for efficient meat production. They reach market weight within a reasonable time frame.

SIZE

This breed is medium to large in size, making it suitable for meat production. The larger size contributes to a higher meat yield per rabbit compared to smaller breeds.

FEED EFFICIENCY

Champagne d'Argent rabbits are known for their good feed conversion, meaning they efficiently convert feed into meat. This can result in lower feed costs and a more sustainable meat production system.

HERITAGE BREED

Champagne d'Argent rabbits have a rich history and are considered a heritage breed. By raising them, you contribute to preserving this breed and its genetic diversity.

✕

AVAILABILITY

Finding quality breeding stock of Champagne d'Argent rabbits in the US may be more challenging compared to more commonly available commercial meat breeds. You may need to invest time and effort into locating reputable breeders or establishing your own breeding program.

MARKET DEMAND

The market demand for Champagne d'Argent rabbit meat may vary depending on the location. It's important to assess the demand and availability of potential buyers or customers before investing in a breeding program.

BREEDING CHALLENGES

Maintaining a healthy breeding program for Champagne d'Argent rabbits requires careful attention to genetic diversity, avoiding inbreeding, and managing potential health issues that can be associated with larger breeds.

LIMITED RECOGNITION

While Champagne d'Argent rabbits have their own loyal following, they may not be as widely recognized or preferred by commercial meat producers compared to certain commercial breeds. This may impact marketability and access to certain markets.

Silver Fox

Silver Fox is the third oldest breed of rabbits developed in the United States. The breed was on the Livestock Conservancy's threatened list because of its dwindling population, but has recently been moved to "recovering" status. Silver Fox rabbits originated in America and while primarily found in North America, their unique standing fur characteristic and meat production qualities have attracted some interest from breeders in other countries as well.

➕

MEAT QUALITY
Silver Fox rabbits are known for their excellent meat quality. The meat is tender, flavorful, and highly regarded by many for its taste.

DUAL-PURPOSE BREED
Silver Fox rabbits are considered a dual-purpose breed, meaning they are suitable for both meat and fur production. This versatility can provide additional income streams for breeders.

TEMPERAMENT
Silver Fox rabbits are known for their docile and calm temperament, making them easier to handle than more skittish breeds.

GENETIC PRESERVATION
By raising Silver Fox rabbits, breeders contribute to the preservation and conservation of this heritage breed, which is important for maintaining genetic diversity in rabbit populations.

✖️

SIZE AND GROWTH RATE
Silver Fox rabbits are generally larger, which means they require more space and feed compared to smaller rabbit breeds. Additionally, their growth rate may be slow compared to commercial meat breeds.

BREEDING CHALLENGES
Maintaining a healthy breeding program for Silver Fox rabbits requires careful attention to genetic diversity, avoiding inbreeding, and managing potential health issues that can be associated with larger breeds.

LIMITED AVAILABILITY
Finding quality breeding stock of Silver Fox rabbits may be more challenging compared to more commonly available commercial meat breeds. Building a sustainable breeding program may require additional effort and research.

SPACE AND MANAGEMENT
Due to their larger size, Silver Fox rabbits may require more space and specialized housing arrangements compared to smaller rabbit breeds. Adequate space and proper management practices are important for their well-being and optimal growth.

It is worth mentioning that individual rabbits within a breed can have variations in characteristics and performance. Additionally, local conditions, management practices, and specific breeding lines can influence the pros and cons mentioned above. Consulting with experienced rabbit breeders or rabbit enthusiasts in your area can provide more detailed insights based on their firsthand experiences.

Special Considerations

When choosing a rabbit breed for sustainable meat production, there are several special considerations, including growth rate, climate sustainability, litter size, etc. Carefully evaluating these factors can help you make informed decisions to establish a successful rabbitry.

CLIMATE AND ENVIRONMENTS

Unless you live in one of the Southern states, rabbit farming changes seasonally. Even though rabbits are one of the easiest types of livestock to raise, some breeds fare better in certain climates than others.

Your rabbit hutches or cages need to be sheltered during all seasons. In summer, your rabbits need sufficient shading to protect them from becoming overheated from extreme temperatures. During winter, your rabbits will need protection from the rain, snow, and wind. Most rabbit hutches have a wooden roof and sides. However, if you are using stackable or hanging wire cages, place a piece of plywood on the top and the sides to block the wind and rain.

During the colder months, you may bunk rabbits together, but never place a male and female that have reached breeding maturity together. Two females may bunk together; they might disagree in the beginning but will rarely harm each other. Never place two mature male rabbits together, as they will fight and cause physical harm to each other. Also, never ever place another rabbit in the cage with a nursing mother, as she will aggressively defend her territory and babies.

Never use plastic to insulate your rabbits' hutches, as rabbits will gnaw at anything that touches their cages.

Be proactive at cleaning your rabbits' housing during the colder months, as fecal matter can stick to your rabbits' feet and freeze. Keep the wire cage bottom clear so the urine and droppings can fall to the ground instead of building up moisture inside of the cage, which can cause frostbite.

A successful rabbitry relies on production, which means your breeding stock is going to need to be crossed regularly to produce babies.

Breeding

As discussed above, selecting the right breeding stock is crucial.

Reproduction

Rabbits have a high reproductive rate, with all domestic breeds capable of breeding naturally. While larger breeds may sometimes be less agile during mating, they do not require artificial insemination. Natural breeding is the standard practice in rabbit production, as rabbit artificial insemination techniques remain specialized and are not commonly used in typical rabbitries.

Litter size

Poor and inconsistent litter sizes or growth rates can destroy your rabbitry.

The breed of the rabbit directly impacts the litter size. Each breed has different genetic characteristics that affect reproductive traits, including litter size. Here are some of my general observations.

Larger breeds, strangely enough, have smaller litter sizes when compared to smaller breeds. For example, Flemish Giants tend to have smaller litter sizes, often four

Photo Courtesy of Kelly Hurley, Phillips Farm

to eight kits for each litter. This may be due to hormonal differences.

Medium-sized breeds have larger litters compared to larger breeds. For example, Californian rabbits, New Zealand Whites, and Satins have an average litter of 6 to 10 kits per litter. For this reason, these breeds are the most popular with commercial breeders for meat production.

Smaller breeds vary in their litter size; some have an average of three to six kits per litter, and others have 8 to 12 kits per litter. Small rabbit breeds are not recommended for meat production due to their growth rate; by the time they reach the ideal weight, they are considered to be roasters or stewers.

Environmental factors, genetics, health, and the age of the doe also play a role in litter size. Proper breeding management, nutrition, and care can contribute to optimal reproductive outcomes.

Growth rate

Look for breeds that have good growth rates and efficient feed conversion. This will help optimize meat production. It's important to provide proper nutrition and a balanced diet to ensure the rabbits reach their full growth potential.

Housing and management

Depending on the breed, housing requirements may vary. Some larger breeds might require more spacious cages or hutches. Ensure proper ventilation, temperature control, and cleanliness for the rabbits' health and well-being.

Care of young rabbits

When it comes to raising rabbit offspring, ensure a clean and safe environment. Provide proper nutrition, including a milk replacement formula if necessary, and monitor their growth closely. Weaning age may vary, but it typically occurs around six to eight weeks.

Culling and selection

Regularly evaluate the offspring for desired traits and cull any rabbits that don't meet the production standards. Selecting the best animals for breeding will help maintain and improve the quality of your meat-producing rabbit stock.

Winter months

During the colder months, rabbits will not "breed like rabbits." Like poultry, they love the longer, sunny days during the summer months and are more inclined to breed then. If you plan on breeding during the darker winter months, you will need to supplement with artificial light until 9:00 or 10:00 p.m.

Expert rabbit farmers plan their batches to pack the freezers during the more agreeable months, so by the time January rolls around, all is going smoothly.

New moms may neglect to pull hair for their kits, or they may give birth on the wire. If you find a batch of unprotected babies outside, bring both the doe and the babies inside. Make a nest using hay or straw and any hair the mother has pulled. If the kits are cold, place them in a warm environment such as near a furnace or woodstove, but avoid direct heat that could overheat them. Once warmed, return them to the nest box with the mother.

While cold weather presents challenges for raising kits, properly prepared nest boxes can successfully protect newborn rabbits even in freezing temperatures. A well-designed nest box with adequate bedding material (like straw) and plenty of fur pulled from the mother will create an insulated environment that keeps kits warm, even when outside temperatures drop below freezing. Many successful rabbitries breed year-round in cold climates with proper nest box management. If you're concerned about extreme cold, providing additional protection such as deeper bedding or moving the does to more sheltered areas during kindling can be beneficial.

Be sure to check the rabbits' hutches frequently, as a kit can easily fall out of the nest, and does rarely look for missing kits and, even if they do, are unable to bring them back to the nest. Use a flashlight around the cage to search for misplaced kits. If you find a kit that is very cold, gently warm it, preferably with contact with human skin. If you find a kit that is slightly chilly, place it back in the nest with its littermates, as the heat of the siblings will warm it back up.

Outcrossing

Many rabbit breeders have several different purebred breeds on hand for outcrossing. Line breeding refers to breeding animals from the same breed, such as two New Zealand purebreds. Outcrossing is breeding two animals that are not closely related, such as a case of crossing a New Zealand with a purebred Californian White.

Line crossing is undeniably one of the best methods for ensuring desirable breeding stock, but caution is needed because inexperienced breeders can cause more harm than good. For example, a professional breeder uses line breeding to lock in desirable attributes, but a novice could accidentally maximize bad traits in the colony. Once this mistake has been made, there is not much one can do except to bring in a whole new set of breeding rabbits and start over.

I generally practice outcrossing. I rely on selection to produce big, fast-growing, healthy-growing rabbits. I am currently using purebred Californian Whites from different breeding stock and crossing them with New Zealand does.

Do not shy away from hard-culling your breeders if you have made the wrong selection. Here are my criteria for my breeding rabbits:

1. The breeding stock must be easy to handle for breeding. By rule of thumb, the pair should mount within 30 seconds of being introduced.
2. The does need to be excellent mothers and care for their kits. They should produce a minimum of eight kits or more per litter.
3. They must come from a healthy genetic line. Remember, you get what you pay for. Healthy parents mean healthy babies.
4. They must have good meat-to-bone density. Ideally, fryers should weigh three to four pounds dressed.

Before you start raising rabbits for meat production, you need to identify your market. As mentioned before, this includes restaurants, wholesalers, custom meat stores, farmers' markets, dog food manufacturers, and so on. Once you have identified your potential buyers, then you can choose which types of rabbit breeds will be best suited for your rabbitry.

Traditionally, rabbit prices drop during fall and winter due to the availability of wild game, making it more challenging to market farm-raised rabbits. In contrast, summer months align with peak farmers' market season, providing an ideal opportunity to sell rabbits when demand is higher and markets are bustling

Most well-established rabbitries recommend setting up long-term clients for a specified weekly or monthly order to ensure constant sales throughout the year. Another option is selling through an experienced supplier for a while until you can establish your own clients.

Once you decide on the market niche you want to focus on, then you will need to market your rabbits properly. For example, if your goal is marketing ecological, sustainable heritage rabbit meat, you will need to first establish a reputation for raising high-quality meat rabbits. You can develop your clientele by educating them about the benefits of heritage rabbit meat versus commercial meat rabbits.

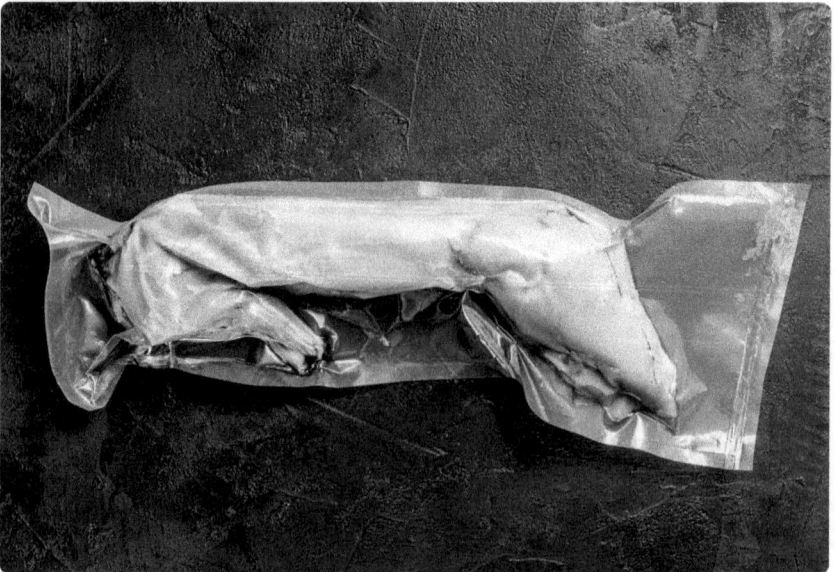

Then you will need to consider the availability of slaughtering facilities according to the regulations of your state or county. Also, once you have narrowed down your potential clients, additional details will need to be researched, such as the type of packaging and transportation costs.

Once you have researched your particular market and chosen the best rabbit breed or breeds for meat production, then you can decide on the size of your operation. The smallest production unit to consider is a herd of around 20 does that are serviced by two bucks.

GROWTH RATE

Profitably raising meat rabbits depends on your breed's average growth rate. Rabbits from strong meat lines give you more meat in less time on less feed. On the other hand, some rabbits take forever to grow and eat a ton of food, which is money out of your pocket.

Not all breeding stock is created equal. Quality breeding stock pays for itself in feed savings alone, as you will have a quicker turnaround time. Only purchase purebred breeding stock from reputable breeders who keep good records. Consider asking the breeder about the kits' weights at eight weeks as a way to gauge the quality of their record-keeping. A breeder who tracks this information carefully is likely to have a well-organized, reliable setup, which can be a great indicator of overall care standards.

This chart indicates the ideal growth rate for commercial rabbit breeds, such as New Zealand Whites, Californians, and Champagne d'Argents.

Meat Rabbit Growth Rate Chart

Age	Normal Growth	Excellent Growth
6 weeks	2.7 lbs	3.25 lbs
8 weeks	4 lbs	4.25 lbs
10 weeks	5 lbs	5.5 lbs
12 weeks	6 lbs	7 lbs
16 weeks	7 lbs	8 lbs

Hybrid vigor is another excellent way to speed up growth rate. As mentioned above, outcrossing is when you cross two different purebred breeds, such as a Californian buck and a New Zealand doe. The first generation of this cross will give what is referred to as hybrid vigor. These kits are going to grow extremely fast and be ready to dress sooner than if you crossed two purebred rabbits of the same breed.

DID YOU KNOW? Crossbred kits from purebred parents are often raised for meat production, as they can exhibit hybrid vigor (heterosis) that benefits growth rates and hardiness. While crossbreds can be excellent for meat production, those interested in developing new lines or breed improvements should understand that F2 crosses (offspring of two crossbred parents) will show much more variability than the F1 generation. Some breeders successfully work with multiple generations of crosses as part of breeding programs for commercial production or breed development, with proper selection for health and production traits at each generation.

TIPS FOR PURCHASING MEAT RABBITS

In the following chapter, we will dig deeper to see how to find high-quality breeding stock. But here are a few tips to help you look over a rabbit to determine if it is healthy.

1. **Examine the entire rabbit and look for signs of defects or underdevelopment.** Perhaps the rabbit is small for its age or underweight. If it looks unhealthy, then the chances of it becoming a good producer are not the best.

2. **The eyes should be bright and clean with no seepage or fresh discharge.** It is normal for rabbits to have pinpoints of dry sleep dirt in the corners of their eyes.

3. **The rabbit's teeth should be even on the top and bottom.** Avoid rabbits with chipped, curled, or overgrown teeth. Rabbits tend to gnaw on anything they can get their teeth on, so it is logical there might be a very small chip.

4. **There should be no discharge from the rabbit's nose.** However, if it is an extremely hot day, the rabbit may have CLEAR discharge, as it is the rabbit's way of sweating. Avoid a rabbit that has mucus.

What breeds do you raise and how did you decide to select them? What advice would you give to someone about selecting the right breed for their farm?

Selecting the right rabbit breed for your farm involves considering several important factors, as shared by our panel of experienced breeders. First, identify the purpose of your rabbits—whether it's meat, pelts, wool, or a mix—since this will guide your breed selection. Breeds like New Zealand and Californian are popular for meat, while Silver Fox and Rex offer dual-purpose benefits for both meat and pelts. Growth rate, meat-to-bone ratio, temperament, climate adaptation, and additional characteristics such as pelt quality or wool production also play significant roles in the decision-making process. By evaluating these factors, you can choose the best breed suited to your needs and environment.

1. PURPOSE OF THE RABBIT

The primary purpose—whether for meat, pelts, wool, or a combination of these—plays a significant role in selecting the breed. Popular meat breeds include New Zealand and Californian rabbits, which are known for their rapid growth rates and high meat-to-bone ratios. Breeds like Silver Fox and Rex also provide dual-purpose benefits, with good pelts as well as meat production.

2. GROWTH RATE AND MEAT-TO-BONE RATIO

Many experts highlighted the importance of selecting a breed that efficiently converts feed into meat.

"Breed what you love! You should have a passion for the breed you choose. I chose Rex for their dual-purpose qualities—best quality fur and, of course, meat. Although they grow a bit slower than other meat breeds, I chose them because their fur quality is amazing and the color variety is fun and they have great temperaments. Raising rabbits should be a pleasure, so do lots of reading about the breeds that interest you."

Kyersten Kerr
Mountain Harvest Rabbitry

"New Zealand rabbits are the best at feed-to-meat conversion, they grow faster on less feed than other breeds. This was an important aspect for me, as I am frugal by nature. I figured if the professionals have this breed, it's good enough for me."

Mia Barcenas
Patch of Heaven
Homestead Farm

New Zealand and Californian rabbits are preferred for their ability to reach butcher weight quickly (about 8-12 weeks). Larger breeds may appear to provide more meat but can have a higher bone ratio, which reduces their efficiency as meat producers.

3. TEMPERAMENT AND HANDLING

The temperament of the breed is crucial for ease of management, especially for beginners. Breeds like the American, Silver Fox, and Rex were noted for their calm and friendly demeanor, making them easier to handle. This is particularly important if the rabbits are part of a family operation involving children.

> "I started out with the American breed because they were a heritage breed in need of conservation help. They have a sweet temperament, which makes them easier to work with. Temperament is key if you're going to be spending time handling them regularly."
>
> **Amy Lambrecht**
> Buckeye Rabbits

4. CLIMATE ADAPTATION

Choosing a breed that is well-adapted to the local climate is essential. For instance, TAMUK rabbits were developed specifically to tolerate high heat, making them suitable for hot climates like Texas. Consideration of climate helps ensure the breed thrives in the conditions they will be raised in.

> "We raise TAMUK-New Zealand crosses. I chose the TAMUKs thanks to their heat tolerance and their ability to keep up with commercial breeds in terms of growth rates and litter sizes when crossed with them. They don't require frozen water bottles or AC in the summer (though they do appreciate it)."
>
> **Sierra Hamilton**
> Crooked Hill Rabbitry

5. ADDITIONAL FACTORS

Other factors such as the aesthetic appeal of the rabbit (e.g., pelt color and quality), heritage conservation, and versatility (e.g., wool production) were mentioned as part of the decision-making process. Breeds like the Silver Fox and Rex were selected for their unique pelts, while Angoras provide wool for spinning in addition to meat.

> "We decided on Silver Fox rabbits because they are a meat breed well known for their calm attitude. They are known to be great mothers and have fairly large litter sizes. This breed has the added benefit of a beautiful pelt that was once sold as an alternative to the pelt of the fox known as the silver fox."
>
> **Stephen Andreanopoulos**
> Quarter Acre Farms

CHAPTER 4

Finding High-Quality Breeding Stock

The key to healthy rabbits is to start raising them the right way—with purebred stock. A purebred rabbit will be pedigreed, but not all pedigreed rabbits are registered. Just because a rabbit is registered does not mean that it will be good breeding stock.

In this chapter, we discuss everything you need to know about finding the highest-quality breeding stock for sustainable meat production.

The Importance of Starting with High-Quality Breeding Stock

Selecting breeding stock for rabbit farming is an important process to ensure productivity and profitability. There are three factors to closely consider when choosing your breeding stock: reproductive soundness, growth rate, and adaptation.

REPRODUCTIVE SOUNDNESS

Selecting a buck with a high libido is especially important for colonies where mating may not always be observed. This ensures that the buck will seek out the does and mate them.

The female's vulva should be well-developed. Small, infantile vulvas should be avoided as they may indicate an underdeveloped reproductive tract. At the same time, evaluate the underlines for teat spacing and quality. Teats should be well-spaced and not too coarse, as they may inhibit the small kits from suckling. Teat number is also important; however, this will vary by breed. A general rule of thumb is the more, the better.

A buck will reach sexual maturity around eight months. Ideally, a buck will maintain its reproductive ability for at least two to three years.

A young male buck should be able to mate one doe at an interval of every three to four days. Bucks more than six years of age should be culled as their semen quality declines.

A female doe becomes capable of reproducing between four to nine months of age, depending on breed size and season. Smaller breeds reach sexual maturity earlier than larger breeds. Does can remain productive breeders well beyond three years of age, with many experienced breeders keeping their proven, healthy does in production for four years or longer. The decision to retain or cull a breeding doe should be based on her continued productivity, mothering ability, and overall health rather than an arbitrary age limit.

Some rabbit farmers prefer purchasing breeding stock that has already been crossed at least once to ensure fertility.

ADAPTION

Some niche rabbit farmers choose to raise heritage breeds because of their adaptive traits. However, many heritage breeds produce kits that take longer to reach market rate, which is why it is recommended to cross a commercial breed with a heritage breed.

Rabbits chosen for breeding stock should be closely observed for mothering ability, litter size, temperament, foraging behaviors, disease resistance, and adaption to different climates. Breeders will keep close records to aid in deciding which animals to keep and which to cull.

GROWTH RATE

I consider growth rate to be one of the most important traits because rabbits that grow faster require less time to process and market. Good record-keeping is the secret to improving the growth rate in your rabbit colony. Size differences can be assessed by visual inspection or with a scale, and the results should be recorded and tracked as the rabbit grows. Any reputable breeder will have a chart or record of an animal's growth rate.

Commercial breeds tend to produce leaner meat than heritage breeds. Heritage breeds originated from a time when the market valued more marbling and fat, so the rabbit farmer should be familiar with the ideal body type for their market. Rabbits chosen for breeding stock should be medium-boned, sufficient to carry their weight, but not heavy-boned, as that reduces yield.

"The very first thing you need to know is how to properly pose commercial breeds. It is easiest to do this on a carpet square, towel, or other non-slick surface. Commercial breeds have a medium length with a depth of body equaling the width of the body throughout. They should have firm flesh and fullness, meaning you shouldn't be able to feel their bones.

To properly pose a commercial rabbit, place their front feet directly under their eyes and align their back feet with the front of the hip. Most of the time, you will use your left hand over the rabbit's head and your right hand to move the rabbit into place. Don't be afraid of moving the rabbit; unless you're using extreme force, you are not going to hurt them.

You will also need to look at their back legs, which means you have to flip the rabbit onto its back. This does take a lot of practice. To flip a rabbit on its back, have it calmly on a secure non-slick surface facing you. Put your dominant hand by the rabbit's head and place its ears between your index and middle fingers. Firmly grab the back of the rabbit's neck and ears with your thumb on the rabbit's forehead as well. Place your non-dominant hand on the hindquarters of the rabbit, then pick up the rabbit enough to slide its back and hindquarters towards yourself onto the stable surface, laying it as flat as possible while still holding the head and ears firmly, and letting them rest their body against your forearm.

It's important to note that if the rabbit does not feel secure, it will flip itself back over. Do not try to stop them, as they can break their backs. Simply let them flip over and start again. If you don't feel comfortable flipping the rabbit yourself, ask the breeder you're buying from to flip the rabbit for you so you can look at their back feet.

Once on its back, you'll be looking for parallel feet that are wide apart, which will give you 'full to the table' and indicate a rabbit that is not pinched in the hindquarters, all equaling more flesh. Should the feet be parallel but narrow, it means the rabbit is undercut. If the feet are not parallel but are wide, it means they are pinched and can lead to difficulty breeding."

Natasha Spudville
D'Argent Rose Rabbitry

Crossbreeds, Pedigrees, and More

As a new breeder focused on meat production, it's helpful to understand the basics of breeding stock and pedigree. While papers can indicate lineage and breed purity, quality rabbits can come from unregistered backgrounds as well, so it's most important to evaluate each rabbit on its health, temperament, and suitability for your goals.

Starting with quality breeding stock—whether purebred or mixed—is key to a successful rabbitry. Purebred rabbits often come with pedigrees or even registration certificates, but if your goal is meat production, papers are secondary to the health and productivity of your animals. For instance, even if a breeder advertises a rabbit as a "Californian," without a pedigree, you won't be able to verify its exact ancestry. However, if the rabbit meets your standards for size, temperament, and health, it may still be a valuable addition to your breeding program.

CROSSBREEDS

Crossbreeds, or rabbits without known lineage, can still make excellent meat stock. While they may lack the predictability of a purebred's characteristics—such as growth rate or meat-to-bone ratio—they are often resilient and adaptable. If you're focused on raising meat rabbits efficiently, don't discount crossbreeds; many experienced farmers achieve great results with mixed lines. However, if consistency is important to you, investing in a few quality purebreds may give you more predictable outcomes in size, temperament, and health.

PEDIGREED AND REGISTERED RABBITS

A pedigree is essentially a family tree, showing three generations of a rabbit's ancestors. For hobby breeders focused on meat production, a pedigree isn't strictly necessary, but it can be helpful if you're looking for consistent traits, like growth rate or temperament. Registration with ARBA adds a level of verification for breed standards, but it's primarily of interest to breeders who focus on show quality rather than meat production. Ultimately, a good meat rabbit doesn't need a pedigree to be valuable; it needs to meet your practical needs and thrive in your setup.

More about pedigrees

As mentioned above, pedigrees provide buyers with all the information required to register the bunny with ARBA if they choose to do so. The pedigree should contain information regarding the rabbit itself and three generations of ancestors, such as parents, grandparents, and great-grandparents. But there is some additional information the pedigree should contain:

Breeder's name	This is optional, but most breeders who sell high-quality breeding stock are proud to stand by their animals.
Rabbit's name	The breeder or the buyer can name their rabbit anything. Breeders often use a prefix before the rabbit's name to indicate the breeder, such as Candi's Smarty, Nelson's Bucky, or Jones's Snowball.
Color	The exact color of the bunny should be written using registration variety. This means instead of simply writing "white," it will be written "blue pointed white," "lilac pointed white," "cream," or something similar.
Ear number	This is the tattoo number that is found in the rabbit's left ear.
Weight	The weight of the rabbit will be written using decimal points, such as 4.5, which in some systems represents four pounds and five ounces, while in others, it is interpreted as four and a half pounds. For clarity, this book follows the convention where 4.5 means four pounds and eight ounces. Only the imperial system will be used, not metric, as is standard in the USA.
Registration and Grand Championship numbers	Provide any ARBA numbers if applicable.

These are all the required details for a pedigree to be registered with ARBA, but most breeders will supply additional information on the pedigree, such as winnings, genotype, ear length, and additional ancestral generations.

A rabbit is considered pedigreed if it has a documented lineage tracing back three generations, regardless of whether the paper has an 'A.R.B.A. Rabbit Pedigree' header or an ARBA member seal. While an official-looking pedigree may add credibility, it is not a requirement for a rabbit to be pedigreed. However, rabbits without a recorded pedigree may have lower value in breeding programs but can still be a viable option for meat production.

More about registration

For a rabbit to be registered with ARBA, the animal must be a purebred, have a pedigree dating back to at least three generations of purebred genealogy, and be at least six months of age. Only then can the rabbit be examined by a licensed ARBA registrar who will certify the specimen as free of physical defects. The registrar will ensure the rabbit meets the minimum physical requirements as established by the breed standard. For example, the rabbit should have the correct size, weight, color, and body type.

The licensed registrar will sign an affidavit to affirm that the animal passed the inspection. An inspection by a licensed registrar typically costs between $6–$10. However, the registrar may charge more for traveling costs.

Each rabbit that successfully passes the examination will receive an official ARBA registration certificate, upon which may be affixed one or more of the following seals:

A **red seal** indicates registration of both parents.

Red and white seals indicate all parents and grandparents are registered with ARBA.

Red, white, and blue seals indicate all rabbits on the registration form were registered with ARBA.

A **gold seal** indicates all ancestors on the registration of the specimen were registered as Grand Champions.

ARBA requires an in-person inspection of the rabbit specimen before granting registration. This inspection is performed by licensed ARBA judges or registrars who assess the rabbit's conformation, breed-specific traits, and adherence to breed standards. This procedure makes rabbit registration one of the best animal registry systems in North America. For example, the American Kennel Club (AKC) does not require an in-person inspection of dogs to provide certification or registration.

The difference in inspection requirements is due to the nature of the animals involved. Rabbits are smaller and easier to transport, making in-person inspections more feasible, while dogs are generally larger and more numerous, making individual inspections impractical for the vast number of registrations.

Please note that specific procedures and policies may change over time, so it's always best to refer to the most current guidelines and requirements from ARBA.

Selecting Quality Stock

Regardless of paperwork, focus on choosing rabbits with desirable traits for meat production—healthy weight, good growth rate, and a calm temperament. An ARBA pedigree or registration seal can be reassuring but isn't essential to produce excellent meat rabbits. As one breeder wisely said, "You can't eat papers." Instead, prioritize animals that demonstrate the qualities you need, whether or not they come with official documentation.

This balanced approach allows you to build a strong, productive rabbitry without feeling obligated to focus on pedigrees and registration unless they align with your long-term breeding goals.

Photo Courtesy of Sierra Hamilton Crooked Hill Rabbitry

How to Find Breeding Stock

Reputable breeders prioritize the health and well-being of their rabbits and maintain clean, well-managed rabbitries. While they take pride in their breeding programs, many do not offer farm tours due to biosecurity concerns that protect their rabbits from disease exposure. Policies on sales and guarantees vary; some breeders may offer limited health guarantees, but refunds or exchanges are not standard practice in the rabbit industry. It is always best to discuss terms with the breeder before purchasing.

Any reputable breeder will not allow you to pick out the rabbits you want, at least not based on sight alone. Personally, I would buy both the doe and the buck from the same breeder, as they know which buck and doe to pair together for success.

The best breeders charge more for their breeding stock and are constantly working hard to upgrade the quality of their rabbits. However, do not expect them to sell you their best rabbits. They will often sell you younger breeding stock or an older breeder that has proven its worth.

Below is a chart of the average prices for purebred breeding stock with pedigree papers for each breed. This chart only provides the average price, as costs can vary from breeder to breeder. Prize-winning rabbits often sell for much more. The general rule of thumb is you should not have to pay much more than the prices cited below unless the animal is extraordinary. On the other hand, I would stay away from rabbits with a cheaper price tag.

RABBIT AGE

A **junior rabbit** is younger than six months of age;
a **senior rabbit** is six months of age or older.
For larger breeds, such as Californian or New Zealand White, a senior rabbit is eight months or older.

NOTE Some breeds like Silver Fox have three age groups — junior, intermediate and senior. These are referred to as 6-class rabbits.

Generally, junior rabbits (under 6 months old) are priced between $20 and $50, while senior rabbits (6 months and older) range from $50 to $100. However, show-quality rabbits or those with exceptional pedigrees can command higher prices, sometimes exceeding $200.

Average Prices Of Breeding Stock

Breed	Junior	Senior
Californian	15	25
Champagne D'Argent	10	25
Checkered Giant	20	30
Chinchilla (American)	15	25
Chinchilla (Standard)	15	25
Flemish Giant	30	70
Florida White	35	60
Harlequin	15	28
New Zealand White	30	45
Rex	23	40
Silver Fox	25	45

RABBIT CLUBS

Rabbit clubs are a fantastic opportunity to determine who has the best breeding stock of your chosen breed. Most rabbit clubs are breed-specific, have established a website or a blog, and often, email weekly newsletters. These newsletters are a gold mine of information for newbies and veterans alike, as they report on point standings from the winners of shows across the country. The leaders of these point standings are the ones you will want to consider purchasing from.

Often these leaders will have their contact information in www.arba.net in the breeder directory. Whether you contact them via telephone or email, be sure to explain your plans for using their rabbits. Most often, they will be thrilled to match you with the best rabbits for your needs.

If you join a local rabbit club, you may find an online group with members who are willing to deliver your rabbit by car at an additional cost.

However, this raises a new problem. Since you are reaching out to some of the best rabbit breeders in the nation, they may not be located in your county. Thankfully, most reputable breeders supply stock throughout the United States via pet transport services or couriers.

RABBIT SHOWS

If you are still on the fence regarding the type of breed to use for meat production or are still looking for a reputable breeder, attend a local rabbit show.

Make sure to see as many different breeds as possible. If possible, have a list of the breeds you are most interested in and write down in advance any specific questions in case you forget. Personally, I like to take a picture of each breed next to something. This helps me scale the size of the adult rabbit, which helps me estimate the size of kits and housing requirements.

The ARBA website often posts the date and location of upcoming shows in each state. Each show listing will have an email and phone number.

Once you have decided on a specific breed, be sure to stay to watch the judging and listen to comments made by the judges regarding each rabbit. This will give you additional insight into disqualifying aspects of the breed.

Be sure to talk with exhibitors and ask where they got their breeding stock and whom they recommend buying from. Often, exhibitors sell foundation stock, and rabbit shows can be a great place to connect with reputable breeders. However, purchasing rabbits at a show requires caution—stress from travel and exposure to other animals can increase health risks. If you are interested in a rabbit from a particular exhibitor, ask about their sales policies and whether a visit to their rabbitry is possible to ensure their rabbits meet your criteria.

Occasionally, I sell my rabbits at shows, but only to experienced farmers who are looking to add a new buck or doe into their colony to improve certain characteristics or try a new breed. Often, the buyer knows my reputation beforehand, and we have chatted previously about my breeding stock.

Here are some different methods for finding a rabbitry that sells good-quality foundation stock.

AUCTIONS

Livestock auctions often sell rabbits for pets or to be processed for meat. Personally, I do not recommend auctions for purchasing quality breeding stock. Auctions often have a no-touch policy, and it is not permitted to inspect the animal closely prior to purchase. Unfortunately, buyer's remorse is common in livestock auctions, as the animals are found to be in poor health.

Some farmers use auctions to pick up supplemental meat rabbits and quarantine them while they finish growing. Then, they process them as fryers when they reach an adequate weight. I find there are too many hidden costs in supplementing my meat production this way, as there are auctioneer fees, additional fees for paperwork, plus transportation costs.

FEED STORES

Local feed stores know all the rabbit breeders in the area. If you are looking for breeding stock in your locality, be sure to visit feed stores and ask. Most feed stores have a bulletin board where you can post an index card with a wanted ad.

INTERNET AND MAIL ORDER

Many breeders connect to potential buyers via the internet. I highly recommend cross-checking these breeders with the official ARBA website. Feel free to email different breeders to inquire about their pedigrees and breeding stock. If you are looking for a breeder in your locality, try searching the rabbit breed of choice and your vicinity. When you find something that looks good, be sure to apply the criteria that you are learning in this chapter.

Most likely, the breeder will ship your rabbits via a specialized rabbit transport service. The shipping charges will vary depending on the number and weight of the rabbits. Rabbits are hardy travelers, and the breeder will ship them with sufficient food and water for their journey.

What should I expect when bringing my rabbits home?

Be sure to have your rabbits' hutches prepared before bringing them to their new home. Make sure you have a supply of clean water, food, and hay. In the following chapter, we will discuss how to properly set up a rabbit hutch.

Once you bring your rabbits home, place them in their hutches and leave them alone so they can get used to their new environment.

How many rabbits should you start with?

Any experienced rabbit farmer will tell you to start small, even if you are planning to have a large-scale rabbitry. Avoid the urge to rush out and buy a whole herd at once; instead, keep it small in the beginning and work out any quirks before the herd grows. I recommend starting with four rabbits, a couple of junior bucks, a junior doe, and a senior bred doe.

The junior bucks and doe should be from top breeders. The senior bred doe should be at least two years old and should have a year or two of breeding life left. Ask to have the senior doe bred to the breeder's best buck. By the time the senior doe gives birth and is ready to mate again, the two junior bucks will be mature and ready to mate, as will the junior doe. So, within three to four months, you will have your original four rabbits, one litter of grow-outs, and two more litters on the way.

How do I build my herd?

Let's say you purchased four rabbits in January. The already pregnant senior doe gives birth in late February. You decide to save the healthiest three does and the best buck from this litter. Later in May, both the senior doe and junior doe give birth. Again, you save three does and the best buck from each litter. By July, you should have bred the senior doe and junior doe again, as well as the three does from the first litter. You will have five litters on the way. By fall, your original four rabbits will have duplicated to over 100. Bear in mind that you will need to have adequate space for all these rabbits.

In Chapter 7 of this book, we will expand on the breeding process. But just to give you a general idea: the average doe bears healthy, large litters for about three years. Bucks can remain fertile for six to nine years, though some breeders choose to retire them earlier based on fertility, productivity, or genetic considerations.

What should I expect upon receiving my rabbits?

If you are picking your rabbits up in person, the breeder will give you the pedigree papers. Each paper will have a number that will correspond with the tattooed number inside the rabbit's ear. If you had your rabbits shipped to you, let the breeder know that you have received them and that they are healthy. Once the breeder receives oral confirmation that you accepted the rabbits, then they will send you the pedigree papers.

If you ordered an impregnated doe, you should also receive the date she was bred and a copy of the pedigree of the buck to whom she was mated.

If you are not fully satisfied with your rabbits, discuss any concerns with the breeder as soon as possible. While some breeders may offer limited guarantees, refunds or exchanges are not standard practice in the rabbit industry. Policies vary, so it is important to clarify terms before purchasing.

ASK THE EXPERTS

How do you source good breeding stock? What tips would you give to a new breeder of selecting the right breeding stock?

Sourcing quality breeding stock is one of the most important steps in starting a successful rabbit farming venture. Our experts highlighted several key points that new breeders should consider when selecting their initial breeding stock, from ensuring the breeder's reputation to evaluating physical and genetic traits. The tips below provide a comprehensive overview, covering the importance of genetic health, meat quality traits, and practical advice for engaging with breeders. By following these guidelines, you'll set a strong foundation for a productive and healthy rabbitry.

1. REPUTATION OF BREEDER

Many respondents emphasized the importance of sourcing breeding stock from reputable breeders rather than auctions or casual backyard breeders. It's advisable to look for breeders who focus on meat production and adhere to specific breeding standards. ARBA (American Rabbit Breeders Association) is often recommended for finding reliable breeders.

"Ask the breeder what they are breeding for and how they choose what to cull. A good breeder will sell the best and eat the rest, selling only what they would themselves keep for breeders. If they're selling the whole litter as potential breeding stock it might be better to find someone else."

Kyersten Kerr
Mountain Harvest Rabbitry

2. PHYSICAL AND GENETIC EVALUATION

Evaluating the condition of the breeding stock is crucial. Experts suggest checking body condition, ensuring the rabbits are healthy, without visible issues like malocclusion, sore hocks, or other defects. Pedigrees are useful, though not essential if the focus is on meat production quality.

"The best rabbits will have a nice arch over their back starting behind the head, wide hips, and a full loin. Bring or borrow a scale to verify their weights. Any breeder that has a problem with you looking them over and weighing them is a red flag."

Heather Riddell-Ide
Riddell-Ide Farm

3. CONFORMATION AND MEAT TRAITS

Conformation is critical for choosing good breeding stock. Selecting rabbits that have the correct muscle structure, wide hips, and a good loin ensures better meat yield. It is suggested that potential breeders attend rabbit shows to learn proper conformation and interact with knowledgeable breeders.

"If meat production is truly your highest goal, pick a meat breed. They will grow to butcher weight the most efficiently. New Zealand and Californian are reliable, but heritage breeds like the American have very good temperaments, which may be a better fit for new breeders."

Amy Lambrecht
Buckeye Rabbits

4. HYBRID VIGOR AND BREEDING PRACTICES

Crossbreeding for hybrid vigor was frequently mentioned as a method to boost the health and productivity of meat rabbits. Starting with a trio (one buck and two does) and selectively breeding based on performance metrics helps build a robust breeding line over time.

5. ASK QUESTIONS AND INSPECT THOROUGHLY

Engaging with breeders by asking lots of questions is a critical strategy. Breeders should be willing to discuss their rabbits' genetics, lineage, and growth rates. Avoid breeders who are unwilling to answer questions or allow inspections, as this is often a red flag.

"Do not be afraid to ask tons of questions. Breeders are usually very passionate about their rabbits and will happily talk to those who are interested. An immediate red flag to me is if someone is not willing to discuss basic information about their rabbits."

Sierra Cofield
Cofield Farms

CHAPTER 5

Rabbit Setup

Taking time before you bring home your rabbits to set up your rabbitry will go a long way to creating a successful operation. Rabbits are defenseless and need a secure rabbitry that will protect them from diseases, parasites, predators and, in the case of bucks, safeguard them from each other. For example, each mature buck will need a space of its own to prevent fighting. Also, your rabbits will depend on you for a healthy source of food and clean water.

Material and Equipment

Gloves
When picking up the younglings at weaning time to decide which ones to keep and which ones to cull, you will quickly realize what sharp nails they have, and they tend to aim for your wrists or forearms. My antidote is a pair of rose gardener gloves with the fingertips snipped off; however, any gloves that protect up to your elbow will do.

Feed scoop and bucket
You will need a feed bucket to safeguard your feed supply. I use a 20- to 30-gallon steel garbage can. You can either purchase an inexpensive feed scoop or make your own by recycling a plastic bottle.

Nail clippers
Most of your rabbits will be processed before they need their nails clipped, but your older bunnies will regularly need their nails clipped, as the longer nails may get caught in the wire mesh of the cage and could injure the rabbit. Any cat nail clipper will work fine.

Brushes

A wire brush with brass or stainless-steel bristles is handy for removing dried manure droppings that may cling to the wire. A brush with nylon bristles is useful for removing excess hair from the inside of the hutch. A slicker brush for dogs or cats can be used on your rabbits to remove dead fur after molting.

Building tools

You will need to occasionally repair your rabbit hutches. Have on hand a pair of wire cutters, J-clip and hog-ring pliers, and a hammer.

Bucket and scrub brush

You will need a bucket and scrub brush for washing and disinfecting hutch floors and feeders, as well as nest boxes.

Propane torch

This is not necessary, but it is a huge time saver to burn off fur on wire hutches. Even though my rabbits don't shed a whole lot, after processing a bunch of roasters, my torch saves me a lot of time in clean-up.

Paint scraper

If you are placing pans or trays under hutches to catch manure, you will need a scraper to scrape off manure that doesn't slide off on its own.

Pitchfork and shovel

Useful for manure removal. A rake and hoe are useful, too.

Wheelbarrow

This is handy for hauling manure to the compost pile or the garden. Also, I use it for moving straw and feed.

Insulation foam boards and plastic foam trays

Again, if you are living in one of the Northern states, you will need to keep your bunnies warm during the winter months by insulating their cages. However, you'll need to make sure it is installed in a way that prevents your rabbits from chewing on it.

Scale

This is ideal for weighing grow-outs or fryers to make sure they are reaching their growth rate goals. I prefer to use hanging scales that have a hook attached to hold a container or even a wire nest box to prevent the bunny from escaping.

Limestone or deodorizer

During the hot summer months, sprinkle limestone or deodorizer under the hutches or pans to reduce the odor. Neutralizing odors will improve the air quality and reduce respiratory issues.

Record sheets

The key to becoming a proficient rabbit farmer comes down to your record-keeping skills. You will need hutch cards to record mating dates and the overall performance of each doe. Stud cards for each buck will provide you with a clear idea of their overall performance. These cards should be placed on the outside of the hopper feeders and be protected with a plastic cover. You can make your own plastic covers or buy them.

You can purchase pedigree forms, production record sheets, and daily, monthly, and annual summary sheets from rabbit supply stores, or you can make your own. Even more conveniently now you are able to track these via computer programs like Everbreed and apps like PanoBreeder.

Skinning knife and hooks

A good-quality skinning knife or boning knife is the ideal size and shape for rabbit processing. Generally, you can find these knives at any kitchen or restaurant supply store. Skinning hooks are made of pleated steel and are designed to hold the carcass in the correct position for dressing.

Tattoo set

If you are planning on entering your rabbits in shows, they will need to have an ID number tattooed in their left ear. And even if you are not going to show your rabbits, tattoos help to identify who's who in the herd.

Location Matters

The location of your rabbitry will affect the health and safety of your colony. Ideally, the rabbitry should be located in a semi-shaded spot that is on high, level ground. You generally want to avoid areas with full-sun exposure during the day if you have the option. Rabbits prefer cooler temperatures.

If you live in an area with varied weather conditions, place the rabbitry in an area that will provide a natural wind block to prevent environmental stress. For example, trees, bushes, and sheds provide a natural shield from cold winter winds.

If you live in a rural setting, you can put your rabbitry wherever you want as long as it is shielded from the elements. However, in residential areas, there are neighbors and municipality zoning restrictions. Most municipalities do not have laws regarding raising rabbits in urban or suburban settings. Instead of asking the town clerk whether it is okay to raise rabbits as livestock, ask for a copy of zoning ordinances. Most laws do not mention rabbits on the list of restricted farm animals.

But even if there are no rules against raising rabbits in an urban or suburban setting, if your rabbitry does not pass the sniff test of nosy neighbors, most likely, the authorities will shut it down upon receiving numerous complaints.

Another key to success is building an attractive and nonobtrusive rabbitry. A rabbitry that is an eyesore can encourage unfounded complaints, regardless of legality or actual sanitary conditions.

If you decide to place your hutches inside a building, be sure to put them over a sand or gravel floor to absorb moisture.

Here are four points to consider in how you arrange your hutches inside of a building, such as a garage, shed, or semi-open enclosure.

Avoid placing the hutches too close to the walls of the enclosure.

If you place your wire hutches against the wall, or even a few inches away, within a short time, the walls will become coated in urine, molted fur, and even caked with manure. Besides becoming a prime breeding ground for parasites, flies, and diseases, your rabbitry will have an unpleasant smell, and the walls will begin to rot or rust. Instead, place your hutches at least three feet from the wall, creating an aisle so you can access the hutches comfortably.

Hang hutches from ceiling joists or rafters.

Personally, I prefer hanging my hutches instead of using hutch legs, which tend to deteriorate over time due to manure and urine. If you plan to place your hutches on a stand or footings, they will need to be cleaned and painted regularly. Several hutches hanging together will be heavy enough not to sway.

Hang hutches back-to-back.

By hanging the hutches back-to-back, you maximize space and prevent the hutches from being too close to the wall. I typically hang the hutches in the middle of the enclosure, creating a walkway all around, allowing for easy access to hutches for cleaning and feeding. Plus, if you decide to install an automatic watering system, it will be less expensive if the hutches are centrally located. Also, if the water is pumped through a heating system in the winter to prevent freezing, it will require less electricity.

Spacing between two or more tiers of hutches.

To ensure proper clearance from the dropping boards above when you plan to stack two or more rows of hutches, create space between the back walls of the lower tiers. An effective method is to utilize smaller hutches for the bottom tier. In my barn, the top-tier hutches measure

30 inches in depth from front to back, whereas the bottom-tier hutches are only 18 inches deep. This configuration leaves a generous two-foot gap, which is more than sufficient.

Alternatively, you could opt for 24-inch or 30-inch-deep hutches for the bottom tier by shifting that tier forward six inches, thereby maintaining a one-foot clearance in either case. These smaller hutches work well for breeding bucks and growing juniors, and the naturally cooler environment on the lower tier is advantageous for bucks.

How to hang hutches.

If you are planning on building a large commercial rabbitry, you should invest in a more sophisticated, engineered suspension system. But if you are planning on having a smaller operation, you can save money and make your own hanging hutches.

MATERIALS

12-gauge wire (available at any hardware or farmer's supply store)

U staples

1 × 6 board (cut to the length of the hutches)

26 × 8 corrugated fiberglass panels

MANURE SCRAPER

6 × 4 piece of corrugated fiberglass

Wooden handle (you can use an old wooden handle from a broomstick or dowel)

EQUIPMENT

Screwdriver

Screws (for scraper handle)

Saw

Measuring tape

How to set up your hanging hutches.

1. Hang the upper tier of hutches using wire suspension.

2. Suspend the lower tier from the one above it, maintaining a 10- to 12-inch gap between the bottom of the top tier and the top of the lower tier.

3. Secure the vertically positioned one-by-six board to the suspension wires by using U staples.

4. Cut corrugated fiberglass panels to the required size using either a saw or tin snips, ensuring a four-inch overhang both in the rear and the front to shield the lower tier from any urine.

5. Place these panels on the board and over the upper part of the lower-tier hutches, ensuring overlap by a corrugation or two as you progress to prevent any leakage onto the cages below.

6. Create slots in the fiberglass panels to allow them to slide around the suspension wires.

CREATE A CUSTOM MANURE SCRAPER

For a custom manure scraper designed to perfectly fit the corrugated dropping boards, start by cutting a six-by-four piece of fiberglass using tin snips. Next, securely attach it to a wooden handle, which can be crafted from a broomstick or any suitable small dowel. The fiberglass in my barn has endured for 40 years, surpassing the longevity of the initial metal shed in which it was housed. It remains resistant to rot and rust and can be easily removed for occasional thorough cleaning with a hose.

Housing Options

Proper housing and good equipment are important for a successful rabbitry. When choosing the best type of house for your rabbits, be sure to consider the rabbits' comfort and ease of handling. The rabbits should have adequate space for growing and exercising. The housing should allow you access to the rabbits in each unit, the feeders, watering bottles, and nesting boxes, and most importantly, be easy to clean.

Rabbit hutches and cages both have their advantages and disadvantages, and the choice between them depends on your specific needs and circumstances. Here are some common questions related to your rabbits' residence.

What size of housing is best?

Bigger is better! A rabbit's home should be at least three to four times the size of a full-grown rabbit when the animal is stretched out. I go by the guideline of a wire hutch with a floor area of three feet by three feet. For giant breeds, add an extra foot of space.

Does confining a rabbit to a hutch rob it of its freedom?

Domesticated rabbits have been raised in hutches for generations and have never experienced the freedom of being in the wild. They do not know what they are missing. It is impossible to deprive them of something that they have never experienced. The truth is if one of your bunnies should escape, it wouldn't even have a clue about what to do and would soon become an easy mark for a predator.

Can I use a wooden hutch?

As mentioned previously, rabbits gnaw almost everything, including wooden frames. Plus, they soak the wooden frame with urine, and droppings pile up in the corners despite daily cleaning. Wooden hutches are unsanitary, damp, and provide an ideal breeding ground for parasites and disease.

What type of hutch do you recommend?

Without a doubt, I suggest a welded wire mesh cage. Depending on the style of cage, some are self-cleaning and only require occasional wire brushing to remove shed fur and periodic cleaning. The droppings and urine fall through the mesh onto the ground or into pans. The cages are well-ventilated, remaining dry and clean.

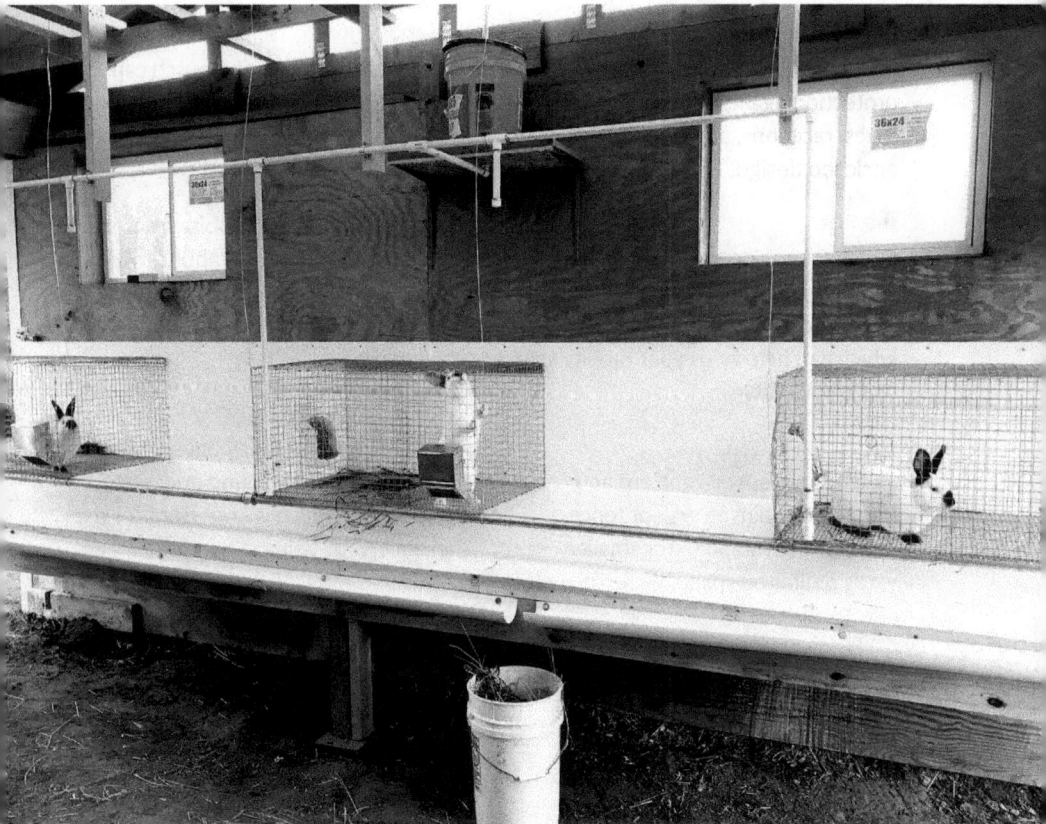

Photo Courtesy of Melina Anderson, Shining Light Farm

Hutches or Cages

Hutches (or cages) are the most popular method of housing rabbits as they are convenient to clean. The structure is designed to protect the rabbits from bad weather and predators. Hutches may be built or purchased prebuilt. They are the best way to ensure the safety of your rabbits, although they might have less room to move around.

As I mentioned, I personally prefer a wire mesh cage because I can control when and with whom I breed my does. Plus, I can keep an accurate record of breeding dates so I can make sure my doe is well-prepared for her litter with a nest box and extra straw. Also, I ensure she gets extra feed while she is lactating. In a colony, this would be impossible, especially if it was a less dominant doe.

Here are some pros and cons of using a hutch or a caged enclosure.

+

✕

PROTECTION FROM PREDATORS
Hutches typically offer better protection against predators like foxes, raccoons, and birds due to their enclosed design.

LIMITED SPACE
The confined space of a hutch can be limiting for rabbits. Rabbits need room to exercise.

WEATHER SHELTER
Hutches provide shelter from adverse weather conditions, including rain, wind, and extreme temperatures, which can help keep your rabbits safe and comfortable.

ISOLATION
Hutches can isolate rabbits from interaction with other rabbits, which may cause the rabbits to become bored or develop destructive behaviors.

SPACE EFFICIENCY
Hutches are space-efficient and can be placed against a wall or fence, making them suitable for smaller yards or limited space.

LESS NATURAL ENVIRONMENT
In a hutch, rabbits may have limited access to natural ground and vegetation, which is important for the animals' well-being.

EASY TO CLEAN
Hutches often have removable trays or floors, which can make cleaning and waste management more manageable.

Colony or Pens

Rabbits require a living area three to four times their size, and in a rabbit colony or pen, this space requirement is easy to achieve. It also provides your bunnies with a space to hop their hearts out and interact with other bunnies.

In the past, most farmers raised their rabbits in a colony setting, just letting them run loose in the barn or a large pen. Often, the rabbits didn't have any organized accommodations, and the farmer would let the bunnies burrow into dirt floors and under haystacks. To some extent, raising rabbits in a colony setting reduced the farmer's workload, as they did not need to clean individual cages.

However, there were many downsides to the colony method. For example, if the farmer wanted rabbit for dinner, he would have to catch it. There was no way of knowing if the rabbit was a tender fryer or a chewy stewer until it was served up on the dinner table. Second, the farmer could not regulate the breeding timing or production, as the rabbits determined that for themselves. Finally, the rabbits were prone to parasitic worms that weakened the colony, often leaving them thin and susceptible to diseases.

Another disadvantage to letting your rabbits live and breed in a colony setting is that they are more prone to predators, such as foxes and weasels, household pets like cats or dogs, and birds of prey. Furthermore, mature bucks will fight and injure each other.

Personally, I find raising meat rabbits in a colony setting to be impractical. For example, does become extremely territorial. After finding several bloody ears and fur everywhere, I decided raising rabbits in a colony setting was not for me.

In a colony, bucks and does can breed indiscriminately, which may be less labor-intensive for the farmer, as the doe chooses the buck herself and takes care of nesting and birthing. However, I found that indiscriminate breeding led to losing more litters, especially during the winter months.

Here are some pros and cons related to raising meat rabbits in a colony or pen setting.

+

✕

MORE SPACE
Pens provide more space for rabbits to move around, stretch, and exercise, promoting their physical and mental health.

SOCIAL INTERACTION
Rabbits in pens often have more interaction with other rabbits, which can lead to better socialization, preventing boredom and obsessive behaviors.

ACCESS TO NATURAL ENVIRONMENT
Pens can be set up in yards or gardens, allowing rabbits access to natural ground, grass, and plants, which is enriching for them.

PREDATOR RISK
Pens may be less secure against predators compared to hutches. Extra precautions are needed to protect rabbits from potential threats.

WEATHER EXPOSURE
In extreme weather conditions, rabbits in pens may need additional shelter or insulation to stay comfortable.

CLEANING CHALLENGES
Keeping pens clean can be more challenging than cleaning hutches, especially if they are not equipped with removable flooring.

In many cases, a combination of hutches and pens can provide the best of both worlds. For instance, you can have a hutch for shelter and security and a larger pen or enclosed run attached to it to provide more space and natural elements. This setup allows your rabbits to have protection and room to roam and play.

I use hutches. The wire caging gives me an opportunity to socialize with my bunnies every day. Plus, I can keep a better eye on the health of the litters. Since my rabbits are used to me picking them up and cleaning out their cages, they are not as territorial when I check on the kits.

We keep each rabbit in an individual hutch or cage, but they still have the opportunity to see, sniff, and touch other rabbits through the wire. Plus, I make sure to give each rabbit personalized attention each day so they do not feel socially isolated.

How to Build a Wire Mesh Cage

Manufacturers of prefabricated cages purchase wire mesh by the truckload, so their costs are often much lower than what an individual would pay in the local hardware store to make their own hutch. I highly recommend shopping around and comparing prices of wire rolls.

Building your own cage is easy. My first attempt to make a cage took me several hours, but now it takes me under an hour. Making a wire cage doesn't require any fancy equipment, just a pair of pliers and wire cutters.

MATERIALS

FOR SIDES
One length of 1 × 2-inch 14-gauge welded galvanized wire fencing (sometimes called turkey wire), 18 inches wide × 11 feet long

FOR FLOORING
One 30 × 36 sheet of welded galvanized wire mesh (not hardware cloth), with a dimension of either 1/2 × 1-inch, 14 gauge or 16-gauge welded galvanized wire mesh (not hardware cloth). It's worth noting that the 14-gauge option is heavier but comes at a higher cost.

FOR ROOF
A single sheet 1 × 2-inch, 14-gauge galvanized wire mesh, 30 × 36 inches

FOR DOOR
A single sheet of 1 × 2-inch, 14-gauge welded galvanized wire mesh, 30 × 14 inches

Latch, dog-leash snap fastener or wire coat hanger

EQUIPMENT

Measuring tape

Hammer

2 × 4 (2 feet long) untreated piece of lumber

J clips or C- ring pliers (approx. 80)

J-clip or C-clip pliers

Wire cutters

Prepare the Side Panels:

1. Place the wire fencing side piece on the ground without cutting it.

2. Utilize a hammer to bend each corner around the two-by-fours' length, forming the two two-and-a-half-foot sides and the two three-foot sides. Ensure you don't bend against the welds.

3. Create a rectangular shape by securing it with J clips or C rings using pliers, spacing them approximately every three inches. If available, J-clip pliers are recommended. You have now assembled all four sides of the hutch.

Prepare the Base and Top:

1. Using pliers and rings or clips, secure the 1/2 × 1-inch floor mesh piece to the sides clips to create the bottom.

2. Similarly, fasten the 1 × 2-inch roof mesh piece to construct the roof using the same method.

Prepare the Entrance:

1. Employ wire cutters to create a one-foot square door opening on one wide side of the door piece, which will serve as the front. Retain half-inch stubs.

2. Safely bend the stubs backward using pliers to ensure there are no sharp edges.

3. The door's wire piece measures 13 by 14-inch to ensure a minimum half-inch overlap on all sides. Attach it securely to the inside of the cage. Many rabbit keepers prefer a top-hinged door, fastened with J clips or C rings, allowing it to swing upward and into the cage. This design aligns with my construction method and is also favored by commercial rabbitries. It ensures that when open, the door remains inside, avoiding interference in the aisle by not extending out, potentially catching sleeves or causing obstructions. Additionally, a top-hinged door provides the benefit of staying closed even if it's accidentally left unlatched, preventing any escape.

Remember, it is easier to care for your rabbits in well-built cages or hutches rather than poorly built, temporary ones.

After the cage, the feed and waterers are the most important elements for having happy, healthy rabbits. However, not all feeders and waterers are created equal.

Some rabbit breeders will argue that you can use an old coffee can for a feeder, but if you look under their hutches, you will find plenty of wasted food. I do not recommend coffee-can dishes as the rabbits can easily tip them over or scratch out their feed, causing it to spill through the wire mesh flooring. Often, rabbits that have these types of feed dishes are underweight because they do not get enough to eat.

I do, however, recommend a self-feeder, and it will not cost you more than $10. A self-feeder attaches to the outside of the hutch, has a lip on the rim of the trough to prevent the rabbits from scratching the feed out, and cannot be tipped over.

Another advantage is that a self-feeder does not take up valuable space on the floor of the hutch, giving the rabbit more space to stretch out its long legs.

WATER REQUIREMENTS

Most adult rabbits drink approximately a quart of water per day. In warmer weather, they will drink more.

WATER AND WATERING SYSTEMS

Providing clean, fresh water to rabbits is essential for their growth and well-being. For many years, I opted for earthenware crocks as the water container for my rabbits. I consciously avoided using tin cans, as they tend to tip over frequently, leaving the rabbits without access to water, and they are prone to rust. While earthenware crocks require regular washing and daily rinsing, they are a better option than tin cans.

Alternatively, you can consider an automatic watering system that connects to a city water supply or a well. This system uses a one- to five-gallon breaker tank to regulate water pressure. Using the force of gravity, the water flows through tubing to reach the valves in the hutches. A semi-automatic system, on the other hand, involves a holding tank, which can be a large jug or a sizable pail, with tubing leading to the hutches. While you need to fill the holding tank yourself, perhaps using a hose, you avoid the hassle of filling individual jugs, bottles, or crocks, and you eliminate the need for constant rinsing and washing. Both types of systems incorporate automatic watering valves.

Automatic watering valves

"Automatic watering systems with drinker nipples save so much time and the rabbits have access to fresh water all the time."

Dave Mengel, Ladybug Acres

An automatic watering valve costs less than earthenware crocks and is designed to be inserted into a pile or a flexible plastic jug or bottle to dispense water from the bottle or tank to the cage.

Valve-jug method

Cut a small hole in a plastic half-gallon bottle or jug using a sharp knife, then insert the valve securely. Seal the opening with epoxy cement to prevent leaks and valve detachment. This method ensures a constant supply of fresh, clean water for your rabbits. With gallon or half-gallon bottles, you only need to refill the water every other day.

Plus, if you're away for a day, there's no need for someone else to water your rabbits. Alternatively, you can purchase premade plastic bottle waterers with attached valves, which may cost slightly more than just buying the valve separately.

Valve-tubing method

This method enhances the performance of the little valve by connecting it to flexible plastic tubing, which is both cost-effective and easy to set up. Even with no prior experience, I successfully created a fully automatic watering system for my rabbits using just scissors and a tape measure.

To set up your tubing system, run the tubing in one continuous line from a one to five-gallon breaker tank or bucket to all the hutches. At each hutch, cut the tubing, insert a tee connector, and attach another piece of tubing with a valve at the end. Secure the valve to the hutch using a bracket and keep the tubing away from the cage wire with standoff clips to prevent the rabbits from damaging it. Finally, add a drain valve at the end of the tubing.

Make sure to source the tubing, valves, and fittings from the same supplier to ensure compatibility. Choose black-plastic tubing to prevent algae growth, as transparent tubing can promote algae even under artificial light.

Winter watering

Below-zero temperature presents a challenge for keeping your rabbits hydrated.

If you are using a crock during the winter, look for ones that have a smaller diameter at the bottom and a wider diameter at the top. Otherwise, if the water freezes, it will cause the crock to crack. In the morning, only fill the crocks halfway. In the evening, dump the water and refill again, and hope the rabbits get a drink before it freezes. If the ice has frozen inside the crock, take the crock inside and place it in a bucket with warm water until it thaws.

If you are using plastic jugs or bottles with valves, you can provide two for each hutch and switch them up in the morning and evening, thawing the other in the meantime.

Semi-automatic or automatic watering systems can keep the water from freezing by using an immersion heater that is placed in a tank of water or a small electric pump that keeps the water moving continuously. Both of these options can be found at your local farmers' supply store and come with instructions on how to install them. However, with this method, there is still the possibility of breakage if the water freezes in the plastic tubing, as the plastic expands as the ice forms.

Another option is to heat your rabbitry using electric heaters or heat lights.

Photo Courtesy of Melina Anderson, Shining Light Farm

Resting Pads and Boxes

Resting pads and boxes provide your bunnies with the ability to rest their paws from standing on wire. Rabbits were not designed to live on wire flooring, and it can cause painful sores on their feet. Unlike cats and dogs, rabbit feet lack protective padding. While wire flooring is often necessary for easy clean-up, it's essential to provide your rabbits with a resting pad or rug for them to sit on.

Some hutches come with a built-in resting section, but if they don't, you can use a piece of rug, plastic mat, or wood. Rabbit-specific resting pads, available from rabbit supply stores, are particularly useful as they're designed with small slots that allow waste to pass through, preventing buildup and keeping the area cleaner. Plastic mats can also be purchased at local farm supply stores. Ideally, the resting area should be level and firm enough for the rabbit to sprawl out while sleeping, and the general rule is the bigger, the better.

Photo Courtesy of Melina Anderson, Shining Light Farm

NEST BOXES

A pregnant doe will need a nest box for her litter. A proper nest box should be able to hold about two inches of pine shavings, shredded newspaper or computer paper, or hay for the doe to kindle (give birth) in.

Approximately 30 days after breeding, the doe will be ready to give birth, so her nest box should be placed inside her hutch about four to five days before her due date. Make sure she is alone in her cage, and all the nesting materials should be new so there is no scent of other rabbits. Once the doe is alone with her nest box, she will arrange the nest to fit her needs.

A typical nest box is a simple wooden box made from scrap wood or plywood (never use treated wood). The measurements can vary in size according to the size of your rabbits. New Zealand Whites or Californians require a box about 18 inches long.

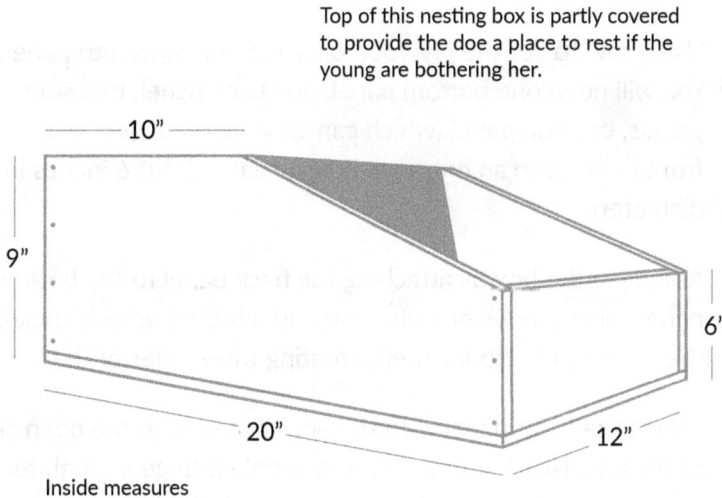

Top of this nesting box is partly covered to provide the doe a place to rest if the young are bothering her.

10"

9"

20"

6"

12"

Inside measures

Here are the materials you will need and step-by-step instructions for making a simple rabbit nest box.

Plywood or sturdy wooden boards

Saw

Screws or nails

Screwdriver or hammer

Measuring tape

Pencil

Hinges (optional)

Steps involved in making your own nest box:

1. Measure and mark the dimensions for the nest box. A common size is about 18 × 18 × 12. Adjust the size based on the size of your rabbits.

2. Use a saw to cut the plywood or wooden boards into panels. You will need one bottom panel, one back panel, two side panels, one top panel (which can serve as a lid), and one front panel with an entrance hole (usually about 6 inches in diameter).

3. Assemble the box by attaching the back panel to the bottom panel using screws or nails. Then, attach the two side panels to the back and bottom panels, creating three sides of the box.

4. Attach the front panel with the entrance hole to the open side of the box. Make sure the entrance hole is large enough for the mother rabbit to enter and exit comfortably.

5. If you want a removable lid for easy access to the nest, attach the top panel using hinges. This will allow you to open the nest box without disturbing the mother and her babies.

6 Ensure the nest box is stable and secure. Sand any rough edges to prevent injury to the rabbits.

7 Place bedding material, such as straw or hay, inside the nest box to create a comfortable and warm environment for the mother rabbit and her kits.

8 Position the nest box in a quiet and sheltered area of the rabbit enclosure where the mother can access it easily.

That's it! You've created a rabbit nest box to provide a safe and cozy space for your rabbits to raise their young.

> **!** **WORD OF CAUTION**
> Never use cardboard for the nest box as it can become wet and break down, or the doe will gnaw away at the box.
> Avoid using built-in nest boxes unless they can easily be removed for cleaning and sterilizing.

Remember to always keep all your nesting cages clean and sanitary, and always provide plenty of clean, fresh feed, water, and nesting material.

WHY ARE NESTING BOXES NECESSARY?

A kit is born hairless, blind, and deaf. A well-constructed nest box will protect defenseless kits from the elements and keep them inside the box until they are big enough to climb in and out by themselves. Another benefit of nesting boxes is that they keep the kits warm, allowing for proper ventilation and preventing the buildup of moisture inside the nest.

ASK THE EXPERTS

What material needs/supplies might a new breeder not realize they need to successfully start raising meat rabbits?

Starting out with meat rabbits requires more supplies than a new breeder might initially realize. While basic equipment like cages, feeders, and waterers are essential, experts recommend investing in good-quality materials to ensure the health and well-being of your rabbits. Beyond the basics, medical supplies, temperature management solutions, and tools for proper record keeping are key components for successful rabbitry management. Additionally, having the right processing tools will make butchering more efficient. The bullet points below provide a comprehensive list of supplies that will help new breeders start off on the right foot, ensuring both rabbit welfare and efficient operations.

1. BASIC HOUSING AND EQUIPMENT

Basic necessities like cages, feeders, and waterers are obvious requirements, but quality matters. Experts emphasized using good-quality wire for cages to avoid injuries and wear on rabbits' feet. Resting mats are recommended for wire floors to prevent sore hocks, especially for breeds like Rex.

2. MEDICAL AND MAINTENANCE SUPPLIES

Many breeders highlighted the need for basic medical supplies, including dewormers, fur mite treatments, and first-aid supplies. Other maintenance supplies such as nail clippers, wire scrapers, and cleaning tools are essential for maintaining the health and cleanliness of the rabbitry.

"You will need a high-quality feed, waterer, and good-quality dust-free grass hay. Other things that I like to promote just because it makes things easier for parasite load and for illness is planting oregano, and giving oregano to your animals when they have the sniffles or if they get sick. Dandelion is also great for vitamins and something the rabbits love."

Jeffrey N B Jenson
High Country Farm

"When it's breeding time, bring the doe over to the buck's cage and watch until three successful fall-offs occur. Then on day 28, bring a nesting box to the doe's cage."

Amberlee Murray
Alpine Rabbitry

3. TEMPERATURE MANAGEMENT AND SHELTER CONSIDERATIONS

Rabbits need protection from extreme temperatures. Proper wind breaks and shade are essential for managing heat, while heated water bottles or other solutions can be necessary during winter. Indoor or well-sheltered setups help manage these conditions more effectively.

4. RECORD KEEPING AND MONITORING

Keeping track of breeding, growth rates, and other data is critical for effective meat production. A notebook, calendar, or app for record keeping, along with a reliable scale for weighing rabbits were suggested by multiple experts.

5. PROCESSING AND MISCELLANEOUS TOOLS

Equipment for dispatching, processing, and storing meat is often overlooked by new breeders. Tools like a sharp knife, a vacuum sealer, and specific items for dispatch (e.g., Hopper Popper) were mentioned as must-haves for efficient meat rabbit production.

"Make sure you have plenty of room and an outlet for rabbits because these things multiply quickly. Get extras of everything. Nail clippers to trim their nails and bottlebrushes if you are using bottles and not an automated system are something easily overlooked."

Nikolaos Karabinas
Karabinas Family Farm

"My top ones are a scale, resting mats if kept in cages, dog nail trimmer, a sturdy surface to evaluate your own rabbits, carrying cages, a tattooing/ID system, a calendar, and a radio. A calendar to write down tasks that need done, when breedings occurred, when does are due, and when grow-outs are born and should be weighed/processed."

Natasha Spudville
D'Argent Rose Rabbitry

Predators

One of the most unpleasant aspects of farming is dealing with predators. Whether they are big or small, the one constant is they can all do serious damage to your rabbit population as well as destroy property. Below are the most common predators you are most likely to find on your farm.

CATS AND DOGS

While it is generally best to keep cats and dogs out of the rabbitry, some experienced breeders successfully manage multi-species environments. Cats and dogs may carry parasites, such as tapeworms, that can be transmitted to rabbits, so proper parasite control and hygiene are essential. Additionally, while some well-trained dogs and barn cats can coexist peacefully with rabbits, their natural predator instincts should not be underestimated. Unsupervised interactions can lead to stress or injury, and even the presence of a predator can cause fear responses in rabbits. To ensure your rabbits' safety, keep feed, utensils, and nest box bedding free from contamination, and introduce other animals to the rabbitry only with caution and proper training.

CHILDREN

Children should be taught how to behave calmly and respectfully around rabbits to avoid causing unnecessary stress. Sudden movements and loud noises can startle rabbits, so it's important to encourage a gentle approach and a quiet voice when inside the rabbitry. Before allowing children in, set clear expectations about handling and interacting with the rabbits. While a rabbitry is not a petting zoo, involving children in rabbit care under supervision can be a valuable learning experience and help foster responsible animal stewardship.

PEOPLE

Unfortunately, there are more and more reports of animal "rescue" groups breaking into small farms and releasing the animals. For this reason, when I am not around, I padlock my rabbitry because I want to protect my bunnies and do not want anyone going in there without my knowledge. If you live in an area where this is common, you may want to invest in padlocks on the hutches and security cameras.

LIVESTOCK

If you have other livestock that share the same barn or living space, it will be a good idea to partition the rabbitry section to keep out wandering livestock, such as chickens. As mentioned before, rabbits need a quiet, calm environment with no mooing, squawking, honking, crowing, or grunting. Goats, chickens, geese, ducks, sheep, pigs, cows, and horses should have their own spaces specifically designed for them.

WILD ANIMALS

Domesticated rabbits' most common predators are opossums, raccoons, skunks, weasels, and badgers. These creatures are experts at breaking into barns since they can squeeze through small spaces and use their paws to rip mesh off the hutches in order to reach your defenseless bunnies. Your best defense against these wild predators is to ensure your rabbitry is securely enclosed with no gaps in the flooring, wallboards, or roof.

Depending on where you live, you may be faced with more predators. Bigger predators, such as bobcats, mountain lions, bears, alligators, wolves, coyotes, and snakes, often require a strong defense. Be sure to enclose your rabbits in a secure enclosure or barn. Use electric fencing, place all trash in animal-proof containers, and store feed in tightly closed containers.

> **WORD OF CAUTION**
> Be sure to check local and federal laws regarding the killing of predatory animals, as in some cases, it may be illegal.

These rodents carry diseases such as hantavirus, leptospirosis, and salmonella. Mice and rats can climb on stored feed bags or even into the hopper feeders, often leaving urine or droppings all over the place. Rats and mice are known to crawl into nesting boxes and kill baby rabbits or even frighten does to the point that they will kill their own litter.

Rodents will sniff out your rabbits' food, and once they find an easy source of food, you will have a hard time getting rid of them. Rats are known to eat rabbit droppings for the extra nutrients. Rodent-proofing the hutches along with regular, thorough removal of leftover food, soiled bedding, and droppings will deter rats and mice. Ensure that the food supply and bedding (straw or hay) are stored in airtight containers.

How to spot a rat

Rodents are experts at hiding in plain sight. Watch out for the following indications that you might have mice or rats:

- Holes in your rabbit hutches
- Teeth marks or other signs of chewing
- Droppings
- An ammonia-like smell (urine)
- Rabbits acting frightened or avoiding a certain part of their enclosure
- Grease marks (Rodents have greasy fur, and as they travel along walls and baseboards, they leave behind a smudge.)

How to keep rodents out of your rabbitry

Rats and mice multiply at an extraordinary rate. For example, a rat's gestation period is just three weeks, and an average litter is 14 rats. Mice can breed almost immediately after giving birth and, in less than 20 days, have a litter of a dozen or more.

However, with the right approach and tools, it is possible to effectively eliminate all types of rodents from your rabbitry and prevent them from returning.

1 Inspect your rabbitry thoroughly.

In order to get rid of rodents, you need to perform an exhaustive inspection to identify how the vermin are getting in and out of your rabbitry. Look around for obvious entry points, such as broken drains, cracks in the walls or doors, and gaps in the foundation. Rats and mice can fit through any opening that your thumb can fit through.

2 Seal cracks and crevices.

Once you have identified any small openings that a rat or mouse can squeeze through, you will need to seal them up, both inside and outside. For best results, fill the cracks or openings with steel wool, metal kick plates, cement, or chalk. Every few months, make sure there are no new openings.

3 Clean up your rabbitry and surroundings.

Rats and mice are drawn to places that provide shelter. Remove yard clutter and move objects away from the walls. If there is anything sitting on the ground that may offer shelter from the elements, chances are it will be occupied by a small visitor.

Substantial vegetation around the rabbitry also presents a risk because rodents can easily hide in tall grass or bushes. Rats and mice are less likely to enter your rabbitry if you make an earnest effort to keep your yard and rabbitry neat and clean.

4 Remove potential food sources.

Removing food sources is one of the best ways to reduce an unwanted rodent population. Mice and rats are not picky eaters and will eat practically anything, including garbage, fecal matter, pet food, and birdseed.

Rabbit pellets are highly nutritious, and all types of rodents find them irresistible. If you notice your rabbits are eating more than normal, it is probably because they are sharing their food with an uninvited guest. Make sure food is stored in containers with lids, and clean up your rabbits' fecal matter daily.

5 Set rat traps.

Rat traps are an effective way to reduce unwanted rodents. I recommend setting plenty of traps in locations with lots of activity. Be sure to use a high-protein, high-carbohydrate snack, such as a piece of cheese, peanut butter, bacon, or buttered bread. Check your traps regularly and replace the bait every twelve hours.

6 Use rat poison or rodenticides.

Rodenticides and poisons that kill off rats and mice by ingestion or absorption through the skin are by far the most effective ways to eliminate mice and rats. Often, these chemicals can be purchased at a hardware store. They should only be used in areas where your rabbits and rabbits' water and food sources will not come into contact with them. And of course, keep these chemicals out of reach of children.

NOTE

Some states have banned the use of certain rodenticides and harsher poisons, so be sure to check your state regulations.

7 Try using deterrents.

There is a wide variety of garden-safe deterrents on the market. These deterrents do not kill the unwanted rodents; instead, they discourage the rodents from spending time in your rabbitry. These deterrents often have a strong scent that repels rodents. You can make your own home deterrents by using vinegar with a few drops of essential oil like peppermint or eucalyptus oil or even fabric softener.

8 Get a cat.

Cats are natural predators of mice and rodents. The mere scent of a cat is enough to encourage unwanted rodents to leave and not come back. However, as noted earlier, be cautious about letting your cat into your rabbitry.

9 Contact a pest control company.

If you have tried the above steps with no luck in eliminating the rodent problem, then you may need to get professional help from a pest control company. They will identify any rodent entry points, locate nests and food storage areas, and eliminate rats without harming your rabbits.

CHAPTER 6

Daily Tasks and Maintenance

A Chinese proverb says, "If you chase two rabbits, both will escape."

Rabbits thrive on routine. This will help you stay organized and avoid forgetting any important daily tasks.

Feeding

Bugs Bunny was not telling us the truth. Rabbits may enjoy snacking on an occasional carrot here and there, but they cannot live on them. Carrots are high in sugar, which can cause tooth decay.

Rabbits will basically eat any type of vegetables or fruit, but if you want to have a successful rabbitry, then you need to feed them what they should eat. Raising meat rabbits means you will need to ensure your rabbits have the best health possible for optimal growth and reproduction, so it is essential you provide them with the highest-quality feed.

PELLETS

Pellets were designed to provide rabbits with everything they need to have a healthy, balanced diet.

A good-quality sack of pellets will specify the protein percentage. I generally use pellets with 16% protein levels as they have a high fiber percentage to prevent digestive issues. Protein content and ingredients vary from brand to brand, but most varieties will keep your rabbits healthy. Be sure to check the ingredient list on the pellet packaging to know precisely which protein sources are used.

How Much Protein Do Rabbits Need?

Dry does, bucks, kits, and developing young rabbits	12–15% protein
	1–4% fat
	20–28% fiber
Pregnant does and lactating does with litters	16% protein
	3–6% fat
	15–20% fiber

Additionally, the protein content in rabbit pellets may vary, so selecting pellets with an appropriate protein level for your rabbits' age and purpose is crucial. The proteins used in rabbit pellets can vary, but they typically include plant-based protein sources. Common protein sources found in rabbit pellets include:

Alfalfa Meal — Alfalfa is a legume that provides a good source of protein for rabbits.

Soybean Meal — Soybean meal is another plant-based protein source often used in rabbit pellets.

Wheat — Wheat can contribute to the protein content of the pellets.

Peas — Some rabbit pellets may contain peas or pea protein as a protein source.

Corn — Corn can also be included in rabbit pellets, although it's more commonly used to enhance carbohydrate content.

Rabbit pellets use plant-based protein sources and are designed to provide rabbits with the necessary nutrients from plant-based ingredients, grains, and legumes.

> **WARNING**
> Never feed your rabbits any foods that contain animal-sourced proteins. Animal-derived proteins are dangerous because rabbits are herbivores, and their digestive systems are adapted for a diet rich in plant materials. Too much animal protein can be challenging for their digestive health, causing illness and even death.

Feeding a rabbit a diet based on pellets has both advantages and disadvantages.

+

NUTRITIONAL BALANCE
Pellets are formulated to provide a balanced diet, ensuring that rabbits receive essential nutrients like protein, fiber, vitamins, and minerals.

CONVENIENCE
Pellets are easy to store, handle, and measure, making them a convenient option for rabbit owners.

REDUCED WASTE
Pellets have minimal waste compared to other diets.

PREDICTABLE NUTRITION
Pellets offer consistency in nutritional content, which can be beneficial for managing rabbit health and growth.

✕

LACK OF VARIETY
A pellet diet alone can be monotonous for rabbits, as they may miss out on the variety of textures and flavors they get from fresh foods and hay.

COST
High-quality pellets can be expensive, especially if you have a large number of rabbits.

DENTAL HEALTH
Rabbits need to chew and grind their teeth regularly, and pellets may not provide the same dental benefits as hay and fresh foods.

OBESITY
Overfeeding rabbits with pellets can lead to obesity, as they are calorie-dense. Proper portion control is crucial.

Personally, I prefer using pellets to feed my rabbits because it is less complicated and gets predictable results. I find it easy to calculate how much feed they need and when. In my opinion, feeding my rabbits a non-pellet diet is full of uncertainties. Plus, my rabbits love their pellets!

I try to strike a balance by incorporating hay, fresh vegetables, and other occasional treats into the diet to ensure my meat rabbits receive a well-rounded and satisfying meal.

SUPPLEMENTAL FOODS

Supplemental foods, such as hay, dry roughage, greens, and concentrates, can help you meet your objectives. However, I strongly discourage using them until you get used to raising rabbits, as it can complicate your feeding schedule and cause health issues.

> **WARNING**
> Never feed fresh greens to your kits as it can give the young rabbits diarrhea, leaving them badly dehydrated, and the consequences may be fatal.

Photo Courtesy of Melina Anderson, Shining Light Farm

Dry roughage

Dry roughage is an excellent source of fiber for your rabbits and includes alfalfa hay, clover, lespedeza, soybeans, and soy hay. Alfalfa hay is my first choice because my rabbits love it. However, rabbits being retained as breeders should receive timothy or orchard hay instead of alfalfa. Alfalfa is often the first ingredient in pelleted feed. By adding alfalfa hay to the breeders diet, you would be giving them too much calcium, which can lead to kidney stones.

Greens

Greens refer to fresh vegetation, such as carrots, rutabagas, sweet potatoes, turnips, and lettuce. Be sure to thoroughly wash all greens before giving them to your rabbits. Only give your rabbits greens that are free from mildew or mold.

Concentrates

Concentrates are dried beet pulp, buckwheat, corn, linseed and peanut meal, oats, and so on. These should be served in their own separate dishes, as rabbits are likely to scratch them out, eating the food they like and wasting what they do not. This method is complicated as it is hard to estimate how much each rabbit consumes.

General Feeding Guidelines

A general rule of thumb is that a healthy adult rabbit should eat approximately one ounce of pellets per pound of its projected adult weight per day. However, this can vary depending on the rabbit's size, age, and environmental conditions. For example, rabbits tend to consume more in colder temperatures to maintain body heat.

Most rabbits finish up their pellets within an hour. You can divide up their pellet portion in 12-hour intervals or feed them once a day. Of course, however you decide to feed your rabbits, you will need to make sure they have a constant supply of fresh hay to munch on.

YOUNG, GROWING RABBITS

Unweaned or just weaned kits should be fed all the pellets they can eat. You may need to return more than twice a day to give them more food. After a while, you will have a general idea of how much food they will need each day. It will vary depending on their age and size.

If you notice that there is food left over from the night before, then give the rabbit a little less. However, if you notice the rabbit devours the food as soon as you serve it, then you need to give it more feed.

MATURE ADULT RABBITS

The pellet consumption for New Zealand meat rabbits can vary depending on age, weight, activity, and other dietary components like hay or fresh greens. As a general guideline, adult meat rabbits typically consume around one-quarter to half a cup of pellets per five pounds of body weight per day.

However, I recommend that you monitor your rabbits' weight and adjust their diet accordingly to maintain their health and optimal growth. Consulting with a veterinarian who specializes in rabbits can provide more precise recommendations based on your specific needs.

PREGNANT OR NURSING DOES

A doe will have increased nutritional needs, especially during the later stages of pregnancy and while nursing her kits. Here are some general guidelines for feeding a pregnant doe.

- **Early pregnancy:** During the first few weeks of pregnancy, you can continue feeding her a regular diet. There's no significant increase in food intake required at this stage.
- **Mid to late pregnancy:** As the pregnancy progresses, gradually increase her daily pellet intake. You might want to increase it by about one-quarter to half a cup of pellets per day, depending on her size and condition.
- **Nursing phase:** Once the kits are born, the doe's nutritional needs will peak. She will need extra protein and calories to produce milk for her babies. You can increase her pellet intake significantly during this time, potentially doubling the amount she was eating before pregnancy.
- **Unlimited hay:** Always provide unlimited access to high-quality hay for pregnant and nursing does. Hay is an essential source of fiber and helps keep the digestive system healthy.
- **Fresh water:** Ensure your doe has access to fresh, clean water at all times.

Remember that individual rabbits may have varying needs, so monitoring weight and condition is crucial. It is also a good idea to consult with a veterinarian experienced in rabbit care for specific recommendations tailored to your doe's unique circumstances.

OBSERVE YOUR RABBITS

Rabbit breeders should watch their colonies closely, especially during feeding time. When you feed your bunnies, run your fingers over each rabbit. If you can feel the rabbit's ribs or spine, then it should be getting more feed. However, if the rabbit is not finishing its pellets, then something is wrong.

REASONS FOR APPETITE LOSS

Thirst

Rabbits will not eat if they are thirsty. Make sure your rabbit has plenty of water. Check the water supply. Is the crock empty? Is the jug's valve plugged?

Boredom

Rabbits are not typically fussy eaters, but in some rare cases they get bored of pellets. If your rabbit seems perfectly healthy and there is no sign of diarrhea under the hutch, then try tempting your bunny with some treats, such as lettuce or uncooked whole oats. This normally will spark your rabbdvit's appetite and get it munching on pellets again.

Overfeeding

Your rabbits should be happy to see you when you come to feed them. However, if they are indifferent, then maybe you are feeding them too much. There are no exact rules, but you will learn over time.

WHEN SHOULD I FEED MY RABBITS?

The best time to feed your rabbits is in the evening as they are the most active then. But if you prefer to feed them in the morning, then your rabbits will adapt to your schedule. If you prefer to feed your rabbits twice a day, use 12-hour intervals in between meals. Whatever you decide, be consistent every day. Never make your rabbits wait to be fed, and never feed them earlier than planned just because you have plans later.

> Most rabbit breeders prefer to feed their rabbits before they sit down for breakfast or dinner.

WHAT ABOUT SALT?

Rabbit pellets already contain all the minerals your bunnies require, so there is no need to feed additional salt to your rabbits.

I discourage using a salt spool, as it tends to drip moisture onto the wire floors, causing them to rust. If you feel your bunnies need additional salt, perhaps because you are giving your rabbits a non-pellet diet, then you can sprinkle some table salt on top of their food using a shaker.

ASK THE EXPERTS

What is the best advice you have for feeding your rabbits? Do you use only commercial foods, or do you grow some of their food yourself?

Feeding rabbits properly is key to their health and growth, and our experts offered a variety of insights into best practices. The foundation of a rabbit's diet often starts with commercial pellets, but many breeders supplement with forage, hay, and even homegrown greens to provide a balanced and varied diet. Seasonal adjustments, careful diet transitions, and considerations about cost and food independence are all important factors in feeding strategies. Below, we summarize the key advice on feeding rabbits to help ensure they stay healthy, grow well, and thrive.

1. COMMERCIAL PELLETS AS THE FOUNDATION

Commercial pellets are a common choice for feeding meat rabbits because they are nutritionally balanced and provide consistency. High-quality pellets with 16-18% protein are recommended to ensure proper growth, health, and nutrition. Many breeders prefer commercial feed for its convenience and assurance of balanced nutrients.

"If you are just starting out with rabbits or getting new breeders I highly suggest using a commercial feed. This cuts down on why the rabbits are not growing right or why they are not breeding. Commercial feed is formulated to have everything a rabbit needs especially in the mineral department."

Natasha Spudville
D'Argent Rose Rabbitry

2. SUPPLEMENTATION WITH FORAGE AND HAY

While commercial pellets form the base diet, many breeders supplement with hay, fresh greens, and forage to improve variety and promote digestive health. Timothy hay and orchard grass are preferred, while alfalfa is often given in moderation as a treat. Some breeders also grow their own greens, such as kale or dandelions, to provide fresh food.

"I use an 18% pellet. The stuff I use is a complete diet and doesn't require hay. I do however sometimes feed hay. You especially need some on hand for a rabbit that might need some extra fiber, or for the babies to chew on. You want a Timothy hay, or orchard grass. Alfalfa can be fed as treats, but not every day."

Amber Irwin
Silver Cloud Rabbitry

3. SEASONAL CONSIDERATIONS

Feeding practices may change with the seasons. In winter, breeders may add rolled oats or black oil sunflower seeds to the diet to help maintain body temperature. During the warmer months, some use rabbit tractors to rotate their rabbits on pasture, allowing them to forage naturally, which can reduce feed costs.

4. IMPORTANCE OF TRANSITIONING DIETS GRADUALLY

Introducing new foods or changing diets must be done gradually to avoid digestive problems. Sudden changes can disrupt gut flora, leading to issues like bloat or stasis, which can be fatal. Gradual introduction helps rabbits adjust to new foods without health complications.

"Please don't overfeed. Rabbit fat doesn't benefit the meat in any way like a good marbling on other animals. If a rabbit eats too much, it can become internally fat, which will make it hard for them to breed as well as the fat typically will encase their vital organs."

Virginia Matter
FluffyButt Ranch

5. FOOD INDEPENDENCE AND COST CONSIDERATIONS

For those aiming for food independence, growing forage and other supplemental foods on their own property is a viable option. However, relying solely on non-commercial feeds may lead to slower growth rates. The choice between commercial feed and homegrown forage often depends on scale, costs, and the breeder's goals for the rabbits.

Watering

Did you know that a five-pound rabbit drinks almost as much water as a 25-pound dog in a day? The average water intake for rabbits is 1.7 to 5.0 ounces per 2.2 pounds of body weight daily. Since math is not one of my stronger skills, it took me a while to figure out how much water my rabbits needed. Here is an estimate of how much water an adult rabbit should drink each day:

Weight: 6 lbs. (2.72 kg)
Water intake: 3.7–11 oz (110–330 mL)

How can you tell if your rabbits are dehydrated?

There are some clear signs to indicate whether your rabbits are drinking enough water:

- Small, dark poop when compared to the rabbit's average-sized and colored droppings.
- Thick, dark, and smelly urine — A rabbit's urine is typically golden yellow without a strong scent. If the rabbit is dehydrated, the urine will be more brownish in color and have a strong odor.
- Lethargy — If a rabbit is dehydrated, it will lack energy and may even start to lose its balance.
- Lack of appetite — Rabbits will not eat if they are dehydrated.

If your rabbit seems to have lost interest in drinking water, there are several things you can do to increase intake.

Give your bunny options.

Rabbits can become bored with drinking from a hanging water bottle or a water bowl. Give your rabbit both options and see which one it prefers. Bunnies tend to like to flip their water bowls. For this reason, I recommend using a crock, a type of ceramic bowl that is heavier.

Provide clean water in a clean container or bowl.

Nobody likes to drink from a dirty glass, so your rabbits shouldn't be expected to have to either. Be sure to empty and refill your rabbits' water sources at least twice a day, more during the warmer months. Additionally, clean the water containers or bowls daily; if not, they can grow a biofilm that may cause digestive issues.

Leave water on their greens.

If you give your rabbits greens to supplement their pellet-based diet, after you wash the greens, re-rinse them just before serving them to your rabbits so they will consume the excess water drops.

Place fragrant greens in the rabbits' water source.

Have you ever had a glass of water with a slice of cucumber? Basically, this is the same idea. Just place some fragrant herbs in your rabbits' water bowl, such as basil, mint, or parsley, and top off with fresh water. This will make the water more appetizing.

Add a few drops of unsweetened juice.

Rabbits definitely have a sweet tooth. Even though it is not recommended to feed your rabbits foods high in sugar, such as carrots, apples, etc., I do recommend adding a few drops of unsweetened carrot, apple, grape, or pear juice to their water just until they get interested in drinking again.

Use bottled or filtered water.

Perhaps your bunny is a water snob. I am blessed that our crisp, clean water is from the Rocky Mountains here in British Columbia, Canada. It tastes amazing, and my rabbits love it; however, I cannot say the same for the water I have tasted in some cities.

Water temperature.

Rabbits are funny little creatures; some like their water chilled, while others like it at room temperature. Play around with water temperature to see if that is the reason for the reduced water intake.

ASK THE EXPERTS

Raising meat rabbits takes a fair amount of work. What surprised you when you first started about the work required to be a rabbit farmer?

Raising meat rabbits comes with more labor than many new breeders initially expect. Our experts shared some of the most surprising aspects of the workload, from breeding challenges to waste management and health maintenance. Understanding these labor requirements upfront can help you plan more effectively and avoid common pitfalls. Below, we summarize the unexpected work involved, with insights on how to manage each aspect to keep your rabbitry productive and sustainable.

1. UNEXPECTED CHALLENGES WITH BREEDING AND BEHAVIOR

Breeding rabbits is not always straightforward. Factors such as temperature, temperament, and timing impact successful breeding. Managing a proper breeding schedule is crucial to maintaining a productive rabbitry, and it can require more effort than anticipated.

2. MANURE AND WASTE MANAGEMENT

Rabbit waste management can be challenging. Waste accumulates quickly, and without an effective plan, it becomes overwhelming. Systems like gutters or using manure as garden fertilizer can make the process more manageable and help prevent issues.

"They poop so much!! They also like to waste food and hay. And they spray; you can get a face full of urine sometimes. You WILL get scratched!"

Amber Irwin
Silver Cloud Rabbitry

3. IMPORTANCE OF EFFICIENT SETUP

A well-designed infrastructure is essential for reducing labor. Properly built housing, efficient waste disposal, and effective feeding systems significantly ease the workload. Taking the time to set up the rabbitry correctly in the beginning leads to a more manageable routine in the long run.

"The amount of daily work required is dependent upon whether or not you built things right the first time."

Heather Riddell-Ide
Riddell-Ide Farm

4. HANDLING AND HEALTH CHECKS

Routine health maintenance, such as nail trimming and health checks, takes more time than expected. Ensuring each rabbit is healthy and keeping up with individual needs adds to the labor, particularly for larger rabbitries.

"How much manure and urine one rabbit can produce certainly surprised me the most. Next would have to be how long it takes to trim everyone's nails and do health checks."

Natasha Spudville
D'Argent Rose Rabbitry

5. PROCESSING AND BUTCHERING

Processing and butchering meat rabbits is labor-intensive, both physically and emotionally. It takes time and practice to become efficient and comfortable with this task. Proper preparation and experience gradually make the process smoother.

"I couldn't believe how long it took me to butcher when I first started out. After 5½ years, I've got my time down to about 14 minutes from bop to bag. Be prepared to be bad at it for a while if you're not already used to butchering, but keep trying and learning knowing that you will get better."

Rachel Heaton
Hardly Simple Farming

Cleaning Pens and Equipment

Now that you have put together your rabbitry, complete with rabbit housing and nest boxes, you will want to keep it in tip-top shape.

Poor sanitation can lead to disease and be a turnoff for neighbors and even potential customers. A clean rabbitry is a key element to raising a healthy and productive herd.

Never place a rabbit into a dirty hutch or cage that was used by another rabbit. Cages can harbor bacteria, parasites, or lingering scents from previous occupants, which can cause stress or health issues for the new rabbit. This is especially important when separating weanlings from their mother and placing them into a separate hutch, as young rabbits are more vulnerable to infections and stress-related illnesses. Cleaning and sanitizing the hutch thoroughly before introducing a new rabbit helps ensure a healthy, comfortable environment and reduces the risk of spreading disease.

HOW TO CLEAN A HUTCH OR A CAGE

1. Using a wire brush, rub off any droppings.
2. Using a torch, burn off the excess fur that you were unable to brush away. Be careful not to torch the wire too long, as it can damage the galvanizing and ultimately cause the cage to rust.
3. You can use a commercial disinfectant according to packaging instructions or make your own by mixing an ounce of bleach with a quart of water. If you have several hutches to clean, consider investing in a tank sprayer. If not, a simple hand sprayer will suffice. I like to spray the hutches with disinfectant and let them dry, and then rinse with plain water and let them dry again before letting the new rabbits go in.
4. While you are washing, disinfecting, and rinsing the cage, do not forget to follow the same steps above for the nest boxes, feeders, and pans. If possible, let these items dry in the sun.

Be sure to regularly clean off hair, cobwebs, and dust that can pile up on top of the hutches, as well as the suspension wires, legs, or any other supporting parts and pieces. Dust is one of the main causes of respiratory problems in rabbits.

Brush hutch floors daily with a wire brush to remove any droppings that have not fallen through the wire flooring. Brushing your rabbits' hutches should be part of your daily routine. It does not take very long, but it will save you a lot of time in the long run. Plus, keeping your rabbits' hutches clean will prevent sore hocks, which develop from having wet feet.

DAILY CLEANING

Remove waste	Scoop up soiled hay (bedding), droppings, and any uneaten food.
Refresh water	Fill bottles or bowls with fresh, clean water.
Check for wet spots	Look for damp or wet areas in the bedding and replace them.

WEEKLY CLEANING

Full hutch cleaning	Once a week, remove your rabbits to a safe, temporary enclosure. Empty the hutch completely.
Scrub and disinfect	Clean the hutch with a pet-safe disinfectant. Make sure to rinse thoroughly to remove any cleaning residue.
Replace bedding	Add fresh, clean bedding material, like hay or straw, to the hutch floor.
Wash food and water dishes	Clean and sanitize food and water dishes.
Inspect for damage	Check for any wear and tear on the hutch structure, including chewed or damaged areas. Repair or replace as necessary.

MONTHLY CLEANING

Deep cleaning	Take the opportunity to do a more thorough cleaning, including washing and disinfecting all accessories, such as toys and hiding spots.
Inspect for pests	Routinely check for signs of pests like fleas and mites and address any infestations promptly.
Check for wet spots	Look for damp or wet areas in the bedding and replace them.

Remember to use a nontoxic, pet-safe cleaning product when disinfecting the hutch. Keeping a clean environment for your rabbits is essential for their health and well-being.

Manure

> *"Rabbit manure is often referred to as 'gold in the garden.' You would be remiss in sustainability if you didn't take advantage of this by either using it in your own garden or selling it to others who will."*
>
> **Heather Riddell-Ide**, Riddell-Ide Farm

Make sure your rabbitry is well-ventilated and, if possible, place the hutches over at least one foot of gravel to provide good drainage. Good drainage under the hutches ensures the ground will stay pretty dry. Wet manure attracts insects such as flies. Plus, it creates ammonia fumes, so it should be removed regularly and not left to sit too long.

Generally, I clean up the manure once a week, mostly because the manure stays dry under the hutches. Often, I will use the manure in my garden or add it to my compost heap. If you are planning on selling the manure, then you will want to pile it in a well-ventilated box until needed.

To store rabbit manure for selling, follow these steps:

1 Collection

Gather rabbit manure from your rabbit cages or hutches. Make sure it's well-composted and free from contaminants.

2 Dry and cure

Spread the manure in a well-ventilated area to dry and cure. This helps reduce moisture content and eliminate any ammonia odors.

3 Screen and sort

Once dry, sift the manure through a screen or mesh to remove any large debris or undecomposed material.

4 Bagging

Package the manure in bags or containers. Seal them to prevent moisture and odors from getting in or out.

5 Labeling

Properly label your manure bags with details like the type of manure, nutrient content, and usage instructions.

6 Storage

Store the bags in a cool, dry place away from direct sunlight. This will help maintain the quality of the manure.

7 Marketing

Advertise your rabbit manure for sale, either locally or online, and ensure you comply with any local regulations or permits for selling organic fertilizer.

> *"Definitely look into raising red wiggler worms under your rabbits. They help keep everything clean and you don't have rabbit droppings everywhere. It's a really neat system, and the worms can also be used for fishing or gardening."*

Kelly Hurley, Phillips Farmm

> *"To be sustainable, it's vital to use every part of the rabbit, including waste. By using their waste, you can enhance your soil and increase your natural food both for human and rabbit consumption. Having a complete plan before you bring the animals in ensures your success."*

Lorie Wood, Sun Fall Farm

Rabbits help your garden. As mentioned at the beginning of this book, rabbit manure is a great natural fertilizer for growing almost everything.

Fresh rabbit manure is 2% nitrogen, 1% potassium, and 1% phosphorous. You can move it straight from under your rabbits' hutch to your garden as it will not burn the plants or soil. You can use your rabbits' pellets on your lawn as mulch for your roses or vegetable garden or even to supercharge your compost pile. An added plus is rabbit manure does not smell as bad as other manure, making it more pleasant to use.

Phosphorus — Rabbit manure has the highest percentage of phosphorus when compared to other types of livestock manure. Phosphorus is crucial for proper plant growth as it aids in transforming sunlight into chemical energy plants need for growing.

Nitrogen — Plants need nitrogen for green growth, helping them reach their full potential in the shortest time. Rabbit manure is great for early-season tomatoes, corn, and salad greens.

Potassium — Potassium, found in rabbit manure, improves yield and quality, and reduces disease that may affect certain vegetables or plants.

Rabbit manure is loaded with a long list of micronutrients and organic matter that improves soil structure, drainage, and moisture retention. Rabbit manure is one of the few natural fertilizers that can be directly applied to your plants without burning them. Plus, there are no undesirable side effects of using rabbit manure on food plants.

Many rabbit farmers have a side business selling their rabbits' waste products. Often, they give the product a quirky name like bunny honey, bunny berries, lawn chocolates, or even Pooh Bunnies and sell it in ice cream buckets. You might not get rich off selling your rabbits' manure, but it will definitely help with some extra costs, such as feed or processing.

How to use it? Just grab a few handfuls and use it as is—either sprinkle it on top of the soil or work it into the topsoil. Rabbit manure may not seem as quick as commercial fertilizers, but in the long run, it will be healthier for your garden by providing food and nourishment for your plants and earthworms.

"Rabbits love attention. Give them attention. If you are consistent with their care, you will notice a problem early and be able to address it before it gets too bad."

Carly Gavlak, Ravine's Edge

Monitoring the health of meat rabbits is essential to ensure their well-being and the quality of the meat they produce. Later in this book, we will go into more depth on many of the points mentioned below. Here are some key steps to monitor your rabbits' health.

- **Regular Observations:** Spend time observing your rabbits daily. Look for any signs of illness or discomfort, such as changes in behavior, posture, or appetite.

- **Check Body Condition:** Monitor the rabbits' body condition. Ensure they have a healthy weight and are not too thin or obese.

- **Clean Environment:** Keep their living environment clean and dry to prevent the spread of disease. Remove waste and replace bedding regularly.

- **Proper Nutrition:** Provide a balanced diet with the right mix of hay, pellets, and fresh water. Ensure the rabbits are eating well and maintaining a good weight.

- **Vaccinations:** Consult with a veterinarian or a rabbit expert to determine if vaccinations are necessary in your area. Some regions require specific vaccinations for rabbits. In Chapter 11, we will discuss different vaccines that may be required in your area.

- **Parasite Control:** Check for signs of parasites, such as fleas or mites. Administer preventive measures as needed, including regular deworming.

- **Regular Health Check-ups:** Schedule periodic check-ups with a veterinarian experienced in rabbit care. They can perform a thorough examination and provide advice on preventive care.

- **Isolation:** Isolate any sick rabbits to prevent the potential spread of disease to the rest of the herd.

- **Quarantine New Additions:** When introducing new rabbits to your herd, quarantine them for a few weeks to ensure they are not carrying diseases.

- **Record Keeping:** Maintain detailed records of each rabbit's health, including vaccinations, treatments, and any observed issues. This can help you track their health over time.

- **Behavioral Changes:** Pay attention to changes in behavior, such as lethargy, aggression, or avoidance of food or water.

- **Respiratory and Digestive Health:** Keep an eye out for signs of respiratory distress or digestive problems, such as a runny nose, sneezing, diarrhea, or constipation.

- **Reproductive Health:** If breeding, monitor the reproductive health of does (female rabbits) to ensure successful pregnancies and healthy litters.

- **Emergency Kit:** Have a basic rabbit health emergency kit on hand, which may include items like wound care supplies, a thermometer, and electrolytes.

- **Learn and Seek Help:** Educate yourself about common rabbit health issues and their treatments. When in doubt, consult a veterinarian or a rabbit specialist for guidance.

Remember that early detection and prevention are crucial in maintaining the health of meat rabbits. Regular care and a keen eye for any signs of illness will help ensure the success of your rabbit farming venture.

ASK THE EXPERTS

What might a new farmer realize about the nitty-gritty details of day-to-day rabbit care?

The day-to-day care of rabbits involves more intricate details than many new farmers expect. Our experts shared insights into what a new breeder might learn about managing rabbit personalities, maintaining hygiene, handling health checks, and dealing with the emotional aspects of rabbit farming. Consistent care and attention to detail are crucial for maintaining a healthy and productive rabbitry. The following points summarize the realities of daily care, helping new farmers prepare for the challenges and responsibilities that come with raising rabbits.

1. PERSONALITY AND HANDLING

Rabbits have distinct personalities, and early, consistent handling helps in developing calmer animals. Gentle handling reduces stress during health checks and processing. Consistent attention also allows for early detection of health issues, such as crusty eyes, sore hocks, or respiratory problems.

> "Rabbits are just like people. They have a wide range of personalities. The earlier and more often they are handled the better their temperament is. No matter how cautious you are, you will get sick or injured rabbits that need tending to right away."
>
> **Jason Lightfoot**
> Deer Run Rabbitry

2. DAILY MAINTENANCE AND HYGIENE

Waste management is a significant part of rabbit care. Rabbits produce large amounts of waste, and keeping the area clean is crucial for their health. Maintaining clean cages, avoiding ammonia buildup, and disposing of droppings properly can help prevent respiratory issues and other health problems.

> "You will be surprised by the amount of waste they produce! Make sure to daily clean trays or buckets, because they will fill up fast!! You may also not realize that their nails need to be trimmed. You may need another person to help you with this, depending on how docile your rabbits are."
>
> **Melina Anderson**
> Shining Light Farm

3. HEARTACHE AND LOSS

Raising rabbits involves challenges, including dealing with illnesses, injuries, and inevitable losses. New breeders must be prepared for potential heartache, such as culling sick rabbits, handling dead kits, or witnessing injuries. This emotional aspect is an unavoidable reality of livestock farming.

"One thing that you might not realize is that there's a lot of potential for heartache. You will get attached to some rabbits and you will have to cull some of those rabbits. Unfortunately, one of the realities of farming livestock is that you will have dead stock at some time."

Angel Wills
Happy Hill Farmstead

4. REGULAR HEALTH CHECKS AND EQUIPMENT

Daily health checks are necessary to ensure rabbits remain healthy. Feeding, watering, and inspecting each rabbit for signs of illness are crucial parts of the routine. Having the right equipment, such as scrapers for cleaning urine buildup and tools for handling emergencies, makes these tasks more manageable.

"I use open crock water bowls because it forces me to pause at every cage and check on each rabbit. Being aware of problems early can avoid many headaches. For instance, you may notice a crusty eye, sore hocks, scratching of the ear, listlessness, grinding teeth, a particular odor of their droppings; all these are signs that you need to address something."

Amy Lambrecht
Buckeye Rabbits

5. WEATHER AND SEASONAL CHALLENGES

Seasonal changes bring specific challenges. Water bowls freeze during winter, while extreme heat in summer can lead to heat stress in rabbits. Planning for these challenges by providing appropriate shelter, managing airflow, and using heating or cooling systems is key to maintaining healthy animals throughout the year.

"Male rabbits can and will 'throw' their urine at you, your children, and each other. Urine guards are your friend, but not a forcefield. Processing a rabbit is not that difficult, but you may want to do it at least once before you invest money into a meat production system and then find out you do not have the stomach for it."

Stephen Andreanopoulos
Quarter Acre Farms

Expanding your rabbitry, or rabbit breeding and raising operation, involves careful planning and consideration. Here are some steps to help you expand:

Research and planning

Research the market demand for rabbits and rabbit products in your area. Understand the costs involved and potential profits. Create a business plan that outlines your expansion goals, budget, and timeline.

Increase housing

Ensure you have adequate and appropriate housing for your rabbits. Construct new cages or hutches or expand existing ones to accommodate more rabbits. Ensure that the housing is clean and safe.

Select healthy breeding stock

As we discussed in previous chapters, your rabbitry depends on the quality of your breeding stock. Choose high-quality breeding rabbits to produce healthy and desirable offspring. Consider the breed you want to focus on and select your breeding stock accordingly.

Feeding and nutrition

Plan for increased feed and water requirements. A well-balanced diet is essential for the health and productivity of your rabbits.

Breeding program

Develop a breeding program to ensure controlled mating, record-keeping, and genetic diversity. Keep track of breeding dates, genetics, and pedigrees.

Health and care

Maintain a strict health and vaccination program for your rabbits. Regular check-ups by a veterinarian are important. Preventive measures can save you from potential disease outbreaks.

Record keeping

Keep detailed records of each rabbit, including birth dates, health history, and breeding history.

Regulations and permits

Ensure you comply with any local regulations, permits, or zoning requirements related to animal breeding and sales.

Scaling gradually

Expanding too quickly can be overwhelming. Consider a gradual expansion to ensure that you can manage the increased workload effectively.

Education

Keep learning about rabbit husbandry, breeding techniques, and the latest industry trends. Join rabbit breeders' associations and networks to gain insights and support.

Financial management

Keep a close eye on your finances. As you expand, you'll have increased costs, so proper financial management is crucial.

Remember that successful rabbitry expansion requires time, effort, and a commitment to animal welfare. Be prepared to invest in your knowledge and resources to make it a sustainable venture.

CHAPTER 7

The Breeding Process

"Take your time and go at your own pace. We were completely new to raising livestock of any kind. We took six months to learn how to care for them properly before we considered ourselves knowledgeable enough to start breeding. Give yourself grace and kindness. You will do everything right and still lose stock, but that's okay."

John and Jessica McCoy, Cajun Hideaway Farmstead

Up to this point, you have carefully selected a breed, obtained foundation stock, and placed your rabbits in their all-new wire cages, and they are now happily munching away on pellets. You have ensured your rabbits have all the comforts and that their needs are being met. But what have your bunnies done for you?

Well, your bunnies have been growing and developing into healthy rabbits. Once they reach breeding age, your rabbits are ready to start producing for you.

Prior to Mating

To produce baby rabbits, you will need to cross your first pair of rabbits. For smaller breeds, the doe and buck should be at least five months of age. For medium breeds, the doe and the buck should be at least six months of age. For larger, giant breeds, the doe and the buck should be at least eight months of age.

However, rabbits do not watch a calendar for permission to breed. There is a reason for the idiom "breeding like rabbits." Rabbits mate so fast that if you blink, you will miss the mounting, mating, and fall-out. It does not take long for a buck to get to work, as reproducing is its main goal in life. Even the youngsters practice their mating skills by riding each other. Basically, a doe can get pregnant as soon as her hormones kick in, but that does not mean she should.

GENERAL RULE OF THUMB
It is best to separate bunnies by 10 weeks of age. I have heard of smaller rabbits getting pregnant at 11 weeks of age. Neglect separating littermates for too long, and you are guaranteed a big surprise. Gender can be determined by two weeks of age, so there are no excuses for not separating them in time.

EXAMINE THE BUCK AND DOE

Once you have decided on the buck and doe you want to mate, you will need to check them out beforehand. Here is how to inspect your rabbits prior to breeding.

Doe

- **Weight** — Weigh the doe to make sure she is a healthy weight for an adult rabbit, according to the breed specifications.
- **Fur** — Examine the condition of the doe's fur. It should have plenty of sheen, with no shedding or bald spots.
- **Vulva** — The vulva should be a reddish color, not a pale pink. The darker the color, the better, as it is a sign of good health and readiness to breed.

Buck

- **Weight** — The weight of the buck is not as important as with the doe, but if he is overweight, he may be too lazy or lethargic to mount the doe.
- **Fur** — Examine the condition of the buck's fur. The shinier the coat, the healthier the rabbit. The buck should not have any bald spots or excessive shedding.
- **Eyes** — The buck's eyes should be bright and alert.
- **Testicles** — The testicles should be completely descended into the scrotum, and the scrotum should be full and large.

If the testicles look withered, are withdrawn into the groin, or there is only one testicle, the buck may be sterile. One testicle may be a temporary condition, as the buck may not have reached sexual maturity, and the second testicle may not have descended into the scrotum yet.

A withered or wrinkled testicle tends to happen to older bucks in colder weather.

Mating

If the prospective parents are in healthy conditions, then you can put the doe into the buck's hutch. Never place the buck inside the doe's hutch, as she can become very defensive about her home. Also, never leave the doe and the buck alone together unsupervised. Once you place the doe in with the buck, you will need to stick around and observe the mating procedure.

Be careful not to blink! Rabbits are fast. If everything goes as planned, the entire procedure will be over by the time you have closed the door to the hutch. The mating process is quick and simple.

1. The buck mounts the doe from the rear.
2. The doe raises her hindquarters.
3. The buck services the doe.
4. The buck falls off the doe, either backward or to the side.

NOTE
Some bucks will also emit a "scream" in addition to a fall-off or even in lieu of one.

Once the buck has fallen off the doe, be sure to remove the doe. Turn her over and examine her vulva to make sure there is a deposit of semen. Return the doe to her cage.

The doe's eggs descend to be fertilized upon mating. This process can take up to eight to ten hours, at which time the eggs are fertilized by the semen and conception takes place. However, if the doe urinates in the meantime, the semen may wash away. For this reason, you will need to repeat the procedure in eight to ten hours because there is really no way to guarantee that the doe was impregnated. I typically mate the buck and the doe two times at 12-hour intervals.

But what should you do if you put the doe in with the buck and nothing happens? Perhaps the buck is not interested in mounting the doe and simply sits in the corner, just ignoring her. Or the buck chases the doe around like crazy, but she resists his every advance.

UNINTERESTED BUCK

If the buck is uninterested, try switching him with another one. I have never had two bucks in a row that were unwilling to service the doe.

But what should you do if you only have one buck?

First, try placing the buck on the doe's back. The buck most likely will get the idea. However, if that does not work, take the buck out of his cage and place him in the doe's cage. Leave the doe in the buck's cage. Leave them inside each other's cages overnight. The buck will pick up the doe's scent, and the next day, he will be more interested. She should be more interested, too, having picked up his scent all night.

If that does not work, then maybe your buck is too fat and lazy. Your best bet would be to put the buck on a diet by reducing his daily food intake. In the meantime, you can attempt mating them again, but before mealtime, so the buck will not be lethargic.

UNINTERESTED DOE

For every uninterested buck, there will be several dozen uninterested does. When a doe is uninterested, she may hunch down in a corner of the buck's hutch. She may attempt to climb the wall to get away from the buck or flatten her body to the floor. Basically, she will resist the buck's every advance by not raising her hindquarters or lifting her tail to let the buck service her.

Often, when you place the doe in with the buck, she will try to toy with him a bit before letting him mount her. Perhaps she will run from him, acting as if she has no interest in mating. But watch her tail. If it twitches, then she is just playing hard to get. A twitching tail is an indication the doe wants to be romanced first.

Sometimes, I restrain the doe for a forced mating. I do this by placing just her rear end inside the buck's hutch. I hold her in place by the scruff of her neck with one hand, and I place my other hand under her belly. Then, I use my fingers to lift the rear end slightly. Normally, the buck will mount the doe and service her. If your buck acts shy, try placing him on top of the doe and keep trying. The more you handle and pet your rabbits, the less likely they will be shy.

WARNING
Never leave the doe and the buck alone together unsupervised! They could start fighting and injure each other. Plus, you will not know for sure if she was serviced.

Common Questions

Here are some common questions related to mating rabbits.

What happens if an underage rabbit becomes pregnant?

It all depends on the doe. Instinct may kick in, and the doe may kindle an excellent litter and care for them carefully. However, the downside is the pregnancy most likely will delay or prevent the doe from attaining a healthy adult weight because the growth energy was expended in sustaining a pregnancy.

In rare cases of does who were bred too early, they may kindle the litter and may even clean the bunnies, only to hop off and forget the babies need to be fed. In this case, you will need to foster or hand-feed the bunnies until they are old enough to fend for themselves. But this does not mean this doe is unfit for a future litter; often, the doe will nurture her next litter perfectly.

How do I know my bunnies are ready to mate?

Bunnies start turning into teenagers around three months of age and will start mounting each other. Bucks will start fighting with one another. Some of this hormonal behavior is to establish dominance, but mostly, it is an indication of increased hormone activity. As mentioned before, it is best if they are not allowed to mate until they are older and more physically developed.

When do female rabbits come into heat?

A doe is considered to be an induced ovulator, which means the act of mating causes her to ovulate within hours. A healthy doe has a hormonal cycle of approximately 18 days—12 to 14 days when she is willing to mate with the buck and four days when she is not interested in mating.

How does the doe's hormonal cycle work?

The doe's cycle starts with five to ten mature eggs (sometimes more) that are ready for fertilization. These eggs are held in fluid that has cell-lined follicles that produce estrogen for approximately 14 days. If the doe is mated, the follicles will rupture within 10 hours, causing the eggs to be released, fertilized, and implanted in the uterus.

If no mating occurs, the follicles and eggs simply deteriorate, and the ovaries produce another set of eggs. During this period, typically four days, the doe will reject any advances made by the buck until the new set of eggs has matured and the follicles start producing estrogen.

Is a two-year-old doe too old to breed?

Rabbits stay fertile for years; however, older first-time does have a tendency to become obese if feed is not controlled. Due to the extra weight, the fat deposits clog the reproductive system, making it more difficult to conceive, and if a doe does conceive, often she has one or two kits. To complicate matters, the kits can grow to gigantic proportions, often getting stuck in the birth canal. In this case, a veterinarian will need to perform a costly C-section to attempt to save the doe and her litter.

Can a rabbit become less fertile if she has not had a litter in a long time?

If the rabbit is healthy and is a healthy weight, she can still get pregnant and kindle a litter. However, if the doe is overweight, the chances of getting a healthy litter from her are not the best. Before attempting to mate your overweight doe, try to get her down to a normal weight according to her breed standard. Does as old as four or five can still be bred, but by that age, their bodies are winding down, and the litter size will be small.

How can I prevent my rabbits from mating?

The best and simplest way to prevent your rabbits from mating is to keep them in separate cages. If you want two rabbits of the opposite sex to live and play together, then your only option is to get one of them spayed or neutered.

I just saw my rabbits mating. Is the doe pregnant?

The chances are high she is if both the rabbits are adults. Be sure to mark the date on the calendar. Give the doe a nest box 28 days from today, and expect a litter of bunnies about three days after that. (And do not forget to separate the doe and the buck.)

Pregnancy

Now that you have mated your rabbits, how do you know for sure if conception has taken place? There are a few ways to tell, but some are more reliable than others.

WEIGHT GAIN

This is the most reliable way to determine if your rabbit is pregnant: weigh your doe before mating and record the date and weight. In two weeks, weigh your rabbit again. If she has gained a pound (medium-sized breed), then she is probably pregnant. Of course, you still cannot be 100% sure.

TEST MATING

A week to 10 days after you first mate the pair, place your doe into the buck's hutch to test her reaction. Chances are, if she is pregnant, she will be uninterested in mating. Often, the doe will growl, hiss, whine, and complain about this encounter. This is a good indication, but still not 100% sure. If the doe is uninterested and has also gained an extra pound, that is a pretty positive indication that she is pregnant.

However, if the doe lets the buck mount and service her, then maybe she is not pregnant. Record the date again and weigh her.

BUT HOW DO YOU KNOW FOR SURE?

The most accurate way to tell if your doe is pregnant is palpation. Two weeks after mating the rabbits, pick up your doe and place her on a flat table. Holding the doe by the scruff of the neck and sliding your other hand under her belly, gently feel around the sides for any signs of life.

Within 10 to 14 days of mating, the young will feel like marbles on both sides of the center of the belly, close to the groin area. If you need practice before doing this test, feel another doe that you know is not pregnant. Palpation takes some practice, but it is the only way to really know if your doe is pregnant.

STILL NOT SURE?

You can feel the marble-shaped youngsters in the doe's belly. Your doe has gained a pound. She vocally protests when placed in a hutch with a buck. And the three and half weeks (28 to 34 days) gestation period has ended. Try throwing a handful or two of straw into the doe's hutch, and if she starts picking it up and making a nest, then you can be 100% sure that she is pregnant.

Photo Courtesy of Angel Wills, Happy Hill Farmstead

Nesting

It's approximately 28 to 34 days from the moment of conception to birth; this is referred to as the gestation period. From experience, I find the majority of litters take 31 days. Typically, you will want to place the nest box inside the doe's hutch on the 27th day, as it gives her time to get used to the box and build her nest inside it. It is best to avoid placing the nest box inside the doe's hutch too soon because she may start using it as a bathroom.

WARM-WEATHER BOXES

When overnight temperatures are above 50°F (10°C), place an inch or two of wood shavings in the bottom of the nest box. I do not recommend using sawdust as it can suffocate the younglings. On top of the wood shavings, place a few handfuls of straw. Be sure to use soft straw and not coarse, stemmy hay. Hay often will be confused as a food and will be eaten. Plus, the coarse hay will not mix with the fur the doe pulls from her chest to insulate her nest.

If the weather is extremely hot, add less straw, as the excess straw may cause the young to hyperventilate. If possible, use an all-wire nest box, as it will provide much-needed ventilation.

COLD-WEATHER BOXES

When temperatures are below 50°F (10°C), place a piece of clean, corrugated cardboard on the floor of the nest box to provide additional insulation. Place three to four inches of wood shavings on top of the cupboard for warmth and absorbency. Then, pack in as much straw as possible and hope the doe will attempt to burrow her young into the straw.

On the 31st day, the doe should kindle her litter in the nest box. However, if it is an extra-large litter, she most likely will kindle on the 30th day. If the litter does not arrive on the 31st day, do not lose hope until the 34th day. After that, it was a false pregnancy, and you should try mating her again.

HOW DOES THE DOE MAKE HER NEST?

The day or night before the doe kindles, she will burrow into the straw and start pulling fur from her underside to build her nest. Once the litter is born, the doe will continue pulling fur from her underside to cover the babies. If the doe does a good job, her litter will survive below-zero temperatures (-18°C).

Some does are reluctant to part with their fur in the colder months, so you may need to help by plucking some fur yourself and placing it on top of the young. If the young are not covered, they may perish, even in above-freezing temperatures.

Photo Courtesy of Angel Wills, Happy Hill Farmstead

The Litter

Once your doe kindles, she will need a few days of peace and rest. Dogs, cats, and children can disrupt your doe's recuperation. Upsetting a doe after kindling can cause her to kill her litter or abandon them, so it is essential that you keep everything calm.

Understandably, you will want to check out the new litter. Depending on where you placed the nest box, you should be able to see directly into it. Around day 31, or soon after, you may notice a big pile of fluffy fur moving ever so slightly as the kits begin to settle in.

Try enticing the new mother to leave the nest by giving her a snack, such as some lettuce, apple slices, half a carrot, or beets. The doe will appreciate the special treat and nutrients at this time. She will be distracted while munching on these yummy snacks, giving you a chance to pull out the nest box with almost no opposition. Carefully remove the box and lift up the fur. Typically, I use a wooden dowel, like a chopstick, to move the fur aside and avoid leaving my scent. Count the babies without picking them up.

Remove any dead babies. If there are more than eight kits, consider the option of fostering the extras to another doe with a smaller litter, though many does are capable of feeding more than eight. Some experienced does can successfully nurse larger litters, even grouping them within the nest box to ensure all are fed. Monitoring the litter closely will help you determine if any additional support is needed.

If there are any runts in the litter, you may decide to reduce the litter size by disposing of these smaller ones. Personally, I let them live. Often, they die on their own in a few days, but other times, they turn out just fine.

A WORD ON FOSTER MOMS

Often, I breed two or more does at the same time, in case I need a foster mom for young from a large litter. It is important that you keep track of any fostered babies' health and growth rates. If the rabbits are a different color, it will be easy to identify which bunny is from a different litter. However, if the bunnies are all the same color, mark one of the foster bunny's ears with ink from your tattoo kit. Gently rub some ink on the ear, and within a few weeks, the ink will fade away.

Most lactating does do not mind fostering young from other mothers, especially if the babies are around the same age as her litter. However, avoid placing a newborn baby in a hutch with larger baby bunnies, as it will never survive in the fight for a nipple.

> **!** **WORD OF CAUTION**
> If you are unsure about how the foster doe will react to the youngster, then rub a bit of vanilla extract above her nose. The scent will distract the doe, and she will not be able to smell anything else. The scent will linger long enough for the youngster to pick up the scent of the new litter.

Hand-nursing orphaned bunnies can be challenging, as many standard veterinary formulas lack the specific nutrients found in a doe's milk. However, it is possible to buy formulas specially made for rabbits, which can improve their chances of survival. Alternatively, goat's milk mixed with a bit of heavy cream can serve as a suitable substitute; many breeders have successfully raised orphaned kits this way. With patience and proper care, hand-feeding can sometimes yield positive results.

There is no need to attempt to sex the newborns until weaning time, as it is almost impossible to tell at this age. My rule of thumb is the less you handle the newborns, the better.

EXTRA FEED FOR THE DOE

Since the litter will depend on the doe for milk, it would be wise to give her some extra feed. Typically, I let the lactating doe munch on as many pellets as she wants each day and give her some green tidbits.

The babies will open their eyes in about 10 days, and in 16 days, they will be bouncing around the cage. In the meantime, keep a close watch on the litter in the nest, make sure they are all healthy, and remove any dead babies. When the bunnies start hopping out of the cage, be sure to remove any green tidbits that are in the cage. Once the litter is three weeks old, remove the nest box if the weather is warmer. Wait four weeks if the weather is colder.

Photo Courtesy of Melina Anderson, Shining Light Farm

Weaning

In the past, litters were weaned at eight weeks, and the doe was then bred again, giving the farmer a maximum of four litters each year per doe. However, due to the higher nutritional value of pellets now, it is possible to maintain the doe's health even if you decide to breed her sooner than eight weeks after kindling.

Typically, I rebreed the doe when the litter is five weeks old, and I wean the young at six or seven weeks (depending on the weather). By doing this, I give the doe at least two weeks without caring for her litter and give her the chance to regain her strength before the next litter. This will give you five to six litters a year per doe. Some large-scale breeders accelerate this timeline even more, often breeding does seven to eight times each year.

I have found that if I wait too long between litters, such as eight weeks after kindling, the doe can gain extra weight, making it harder for her to conceive. If you are not planning to rebreed, you don't need to wean the bunnies from their mother right away. However, keep in mind that young bucks can start attempting to breed as early as 8 to 10 weeks, so it's safest to separate them by then to prevent any accidental breeding with their dam. In most cases, mothers and daughters get along well and can be left together for a few months.

Troubleshooting Breeding Issues

Breeding rabbits is a fickle business. Below are common and not-so-common issues you may face, along with some suggestions that will save you money in the long run.

WEATHER-RELATED PROBLEMS

The long-term goal of most rabbit farmers is to have a sustainable rabbitry that produces a steady income year-round. This means you will need to breed your rabbits in all types of weather. But what challenges do breeders face in the hotter or colder months?

Hot weather

Breeders fear they will lose babies to hyperventilation.

Solution? A wooden nest box can become damp and prevent proper ventilation. To prevent this, you can use cooling nests made from wire screening. The wire screening allows for plenty of ventilation. A well-ventilated nest box allows you to breed your rabbits through the hottest months of summer.

Cold weather

Breeders fear they will lose baby rabbits to hyperthermia.

Solution? I prefer to use a wooden nest box, but you could use the same wire nest box from summer and place a specialized electric heating pad in it. These electric pads are designed to fit inside a rabbit's nest box, and if they save just one litter, they will have paid for themselves.

Here are some other ways to keep your litter toasty warm during the winter months.

An aluminum photo-reflector light can be placed over the hutch where the nest box is located. The heat from the bulb will penetrate the nest box, keeping the babies warm. You can also place the light under the hutch, and the youngsters will burrow down into the straw to stay warm.

Ceramic infrared heater bulbs, often used for reptile tanks, can be used in a similar fashion as the aluminum photo-reflector light bulb.

These two heating options are only necessary on the day or night of kindling and for two to three days after, as the bunnies will grow fur and become better able to withstand the cold. With proper equipment, there is no reason why you cannot breed your bunnies year-round.

> *"A lot of breeders I know, including myself, shelve their nest boxes during winter. I know some that do it year-round actually. Shelving consists of removing the nest box from the dam's cage. Then return the nest box to the dam two times per day, once in the early morning and once late at night. Some bring the dam to the nest box as well."*
>
> **Lorie Wood**, Sun Fall Farm

MATING BUCKS

One of the most frustrating problems you will encounter when raising meat rabbits is breeding problems. We have already discussed what to do when the doe is uninterested in mating but not when the buck is uninterested.

Here are some suggestions if your buck does not oblige to service the doe:

- Check to make sure the rabbit is a buck.
- Is he old enough yet? While bucks can breed from 12 weeks, most are not interested until they are at least six months old.
- Is the breeding space small enough to let him catch the doe? Some lady bunnies like to play hard to get.
- Add some apple cider vinegar to his water, and add some ginger, parsley, or raspberry leaves to the buck's diet to boost his libido.

Bucks vary from male to male. For this reason, it is important to keep records of performance results. This will help you select bucks based on their breeding potential. Young bucks can be bred from five to six months old; however, they should be mated gradually, giving you an opportunity to observe performance.

I recommend that the buck only be mated once a day, often waiting 12 to 24 hours before mating again. Usually, the second ejaculation contains much more sperm than the first.

A buck beyond six years or older should not be used for breeding and should be culled, as semen quality declines. Also, if the buck is overweight, he could lack libido, have reduced semen volume, and have poor-quality spermatozoa.

CANNIBALISM

Unfortunately, for a number of reasons, such as nervousness, severe heat or cold, or neglect (failure to nurse), does may kill their young. Cats, dogs, children, or predators can cause a nervous doe to kill and eat her young. For this reason, you should restrict all animals and visitors from entering or roaming near the rabbitry.

Make sure the doe has an adequate supply of clean, cool, and fresh water. Use veterinary-approved electrolytes in the drinking water during the hotter months. Adding electrolytes increases water consumption. If you use a water system to supply your rabbits with water, be sure to flush out the hot, stagnant water at least once a week. If the doe does not drink enough water, she may kill and eat her young.

Cannibalism is a natural instinct of nest-cleaning. However, if the doe kills and eats two litters in a row, then she should not be used for breeding.

How can you tell if the doe ate her young? Some common symptoms are:

- There are missing kits.
- The nest is full of blood.
- Pieces of kits are found.
- Kits are missing limbs or ears or have other body lacerations.

Keeping Breeding Records

It is essential that you keep good breeding records; otherwise, you will have no way of knowing when to place the nest box inside the doe's hutch or when to expect the litter. This is a recipe for disaster. For this reason, you need to record the dates, the buck who serviced the doe, and any other relevant information.

Many breeders use computer programs to streamline their record-keeping, as it makes tracking breeding schedules, health records, and pedigrees much easier. Programs like Everbreed, Kintraks, and PawPRINT Genetics offer digital tools specifically tailored for rabbitry management, allowing breeders to access information at a glance and maintain organized, up-to-date records.

However, keeping records by hand also has its benefits; manual record-keeping provides a tangible backup and allows you to easily track individual observations in real time. For this reason, we'll cover detailed methods for hand-recording in this book, so you can choose the system that best suits your needs.

You can make your own record cards or purchase them at your local farmers' supply store.

THE DOE'S RECORD CARD

The following information should be found on the doe's hutch record card, as it will indicate whether she is old enough to be bred and which buck should service her:

- Birth date
- Name
- Ear identification number
- Names of the doe's sire and dam
- Registration number, if applicable

Whenever you mate the doe, write the name of the buck down and mark the day and time she was serviced. If you test-mate her in 10 to 14 days and she mates, then mark the additional date. However, still respect the first date by placing the nest box in her hutch on the 27th day. If the doe does not kindle within a few days, remove the nest box and return it on the 27th day of the second mating date.

Without a doubt, the doe's record card is by far the most important document for breeding your rabbits successfully. Faithfully recording any information relevant to the pregnancy and the young will help you decide how to mate your rabbits in the future. If the record shows a healthy, large litter from the doe to a certain buck, then you will most likely want to breed these two again. If not, you may want to switch the doe or the buck for another or even get rid of the doe.

Here is some other information that should be found on the doe's hutch record card:

- Name of the buck who serviced her
- Date of service
- Pregnancy check
- Date kindled
- Number of young that were kindled and survived
- Average weight of young at eight weeks of age
- A space for any additional remarks

THE BUCK'S RECORD CARD

If you have a number of bucks in your rabbitry used for breeding purposes, then keeping a record card will provide insight into which buck to use.

The following information should be found on the buck's hutch record card.

- Birth date
- Name
- Ear identification number
- Names of the buck's sire and dam
- Registration number, if applicable
- Name of the doe serviced
- Date of service
- Quality of the litter (excellent, average, poor)
- Number of young kindled and how many survived

Typically, I keep all the records in a plastic baggie on the outside of the hutch so that I do not need to retrieve it from a box or folder. This lets me check up on production quickly to determine how many offspring they are producing. Larger commercial operations often use computerized record cards for all their bucks and does.

> *"Rabbits have been known to chew/eat tags. To avoid this, metal tag holders can be purchased to attach to the cages. Then the information card can be slid into it, preventing the rabbits from reaching it."*

Lorie Wood, Sun Fall Farm

STUD RECORD CARD

Breed _____ Ear No. _____ Reg. No. _____ Born _____

Sire _____ Reg. No. _____ Dam _____ Reg. No. _____

Doe Served	Date of Service	Date Palpated	Date Kindled	Number of Young		Wt. of Litter		Quality of Litter		
				Kindled	Survived	21 Days	56 Days	21 Days	56 Days	Exc	Avg.	Poor

DOE HUTCH RECORD CARD

Doe _____ Ear Tattoo No. _____ Breed _____ Normal Weight _____

No. Nipples _____ Born _____ Sire _____ Dam _____

Buck	Date of Service	Pregnancy Check	Date Kindled	Number of Young		At 8 Weeks		Remarks
				Kindled	Survived	No. Left	Weight	

Sunlight and Breeding

In the Western Hemisphere, the longest day of the year is June 21, and increased daylight plays a role in stimulating reproductive activity in many animals, including rabbits. While some believe warmer spring weather encourages breeding, research has shown that light exposure influences reproductive hormones. Rabbits, like many mammals, benefit from adequate lighting for optimal fertility.

Veteran rabbit breeders note that both bucks and does need sufficient light to maintain reproductive activity. If you are building a rabbitry, consider incorporating large windows or skylights to maximize natural light. If natural light is limited, full-spectrum artificial lighting can help regulate breeding cycles. However, in non-climate-controlled rabbitries, summer heat can reduce fertility, so seasonal temperatures should also be considered when planning breeding schedules.

For example, egg farmers often provide artificial lighting for 16 hours a day to maintain consistent production. Some rabbit breeders use similar lighting strategies to encourage year-round fertility, though it's essential to balance artificial lighting with natural seasonal conditions for the health of the animals.

THE SCIENCE BEHIND LIGHT

Light enters through the eye's iris and transmits to the pineal gland, which then sends a signal to stimulate the pituitary, thyroid, and thymus glands. Thyroid deficiency has a direct effect on ovulation.

The science behind light and breeding rabbits involves managing their exposure to light to regulate reproductive cycles. Light affects the production of hormones like melatonin, influencing breeding behavior. Controlling light duration can simulate seasonal changes, prompting rabbits to breed.

WHY FULL-SPECTRUM LIGHTING

Full-spectrum lighting refers to artificial light sources that mimic the entire electromagnetic spectrum of natural sunlight. These lights emit a balanced spectrum of visible light, including all colors, as well as wavelengths beyond what the human eye can see, such as ultraviolet (UV) and infrared.

Full-spectrum lighting is often used in various applications, including indoor gardening, photography, and animal husbandry. For rabbits, exposure to full-spectrum lighting can be beneficial, as it closely replicates natural sunlight. This type of lighting can support their overall well-being, including mood, behavior, and physiological processes influenced by light, such as vitamin D synthesis.

If your rabbits are in a barn or a shed, be sure to put in full-spectrum lighting, especially for the fall and winter months. You can easily purchase these types of bulbs at most feed stores, garden centers, and lighting stores. Full-spectrum bulbs cost more than fluorescent bulbs but trust me; your rabbitry will flourish under superior lighting.

Photo Courtesy of Melina Anderson, Shining Light Farm

Summary

Here is a brief summary of the reproductive stages in a rabbit.

Breeding bucks required	1 buck for 10 does
Age at which they can breed	Small breeds — Five months Medium breeds — Six months Large breeds — Eight-Nine months
Reproduction characteristics	Polyestrous. Does appear to have no definite estrus cycle, although an established rhythm exists in their sexual reproduction. The cycle consists of 14 fertile days and 4 infertile days.
Signs of fertility	Congested, purple, moist vulva; restlessness; rubbing chin on the side of the cage; lying in mating posture with tail lifted.
Mating behavior	The doe is always taken to the buck's cage for mating. If the doe is receptive to the buck's advances, she will lift her tail, and within a minute, the buck will mate the doe. Mating is successful when the buck falls off the doe backward or sideways after mating.
Ovulation	Ovulation occurs 10 to 12 hours after copulation (reflex ovulation).
Gestation period	28 to 34 days (average 31 days)
Pregnancy diagnosis	The methods adopted to determine positive pregnancy are mating, weight gain, and the palpation technique. The palpation technique is the most reliable method if done by an experienced rabbit breeder.
Litter size	Six to eight kits
Weaning	Four to six weeks
Kindling interval	Two months (it may be as short as one month if the doe is to be bred immediately following kindling).

CHAPTER 8

Handling Your Rabbits

Sexing

It can be tricky to tell if your bunnies are male or female. Sexing your rabbits avoids unwanted pregnancies and fighting between males. In this section we will discuss how to hold your rabbit to check its sex and how to tell if it's male or female.

HOW TO HOLD YOUR RABBIT

Make sure you are in a safe and quiet space so that the rabbit does not become easily frightened. Most young rabbits are not used to being held yet, and this will be stressful for them. Often, they will want to try and escape. For this reason, you should gently and firmly hold your rabbit. Make sure you are in an enclosed space in case your bunny manages to wiggle out of your grip.

Personally, I find it easiest to sex my rabbits by sitting and holding them in my lap or on a table. If using a table, put down a towel first to make it less slippery. Never hold the rabbit on its back, as this is extremely uncomfortable and stressful for the animal.

Never roughly force the rabbit into the desired position. Instead, be slow and calm. Ask a family member or friend to help you, as it is not easy to hold the bunny and sex it simultaneously.

1. To hold your rabbit, place one hand under the bottom. This will fully support the rabbit's weight.
2. Place your other hand across the chest.
3. Hold the rabbit upright as if in a sitting position, with its rear paws facing away from you. Now the rabbit is in a position where you can easily identify its sex.

HOW TO TELL IF YOUR RABBIT IS MALE

A young male between eight to ten weeks will have testes. The younger the rabbit, the more difficult these will be to observe. Since you may be touching the rabbit's genitalia, be sure to disinfect your hands between rabbits and wear protective gloves to prevent germ transfer.

1. Hold the rabbit in the position stated above.
2. To see the genitals, part the fur.
3. Gently apply pressure, and the penis should emerge.
4. The anus sits at the lowest point below the genitals and is separate from the genitals.

HOW TO TELL IF YOUR RABBIT IS FEMALE

A young female can be identified around eight weeks of age.

1. Hold the rabbit in the position stated above.
2. To see the genitals more clearly, part the fur.
3. Gently apply pressure, and nothing should emerge.
4. The vagina and anus are immediately next to each other.

Once you know the sex of the kits, place the rabbits into corresponding hutches. If you are unsure if the rabbit is male or female, try again in a day or two to get a more accurate answer. However, you may want to keep the rabbit in a separate enclosure to avoid an unwanted pregnancy.

Ear Identification

Once the rabbits are weaned, they should have a permanent number tattooed in their left ear. But why do meat rabbits need to get an identification number if they are only going to be processed in a few months?

- It helps identify the rabbit.
- It protects you from theft.
- It denotes the rabbit owner.
- It helps indicate the age of the rabbit.
- It's a requirement if you decide to enter your rabbit in a show or register it with ARBA.

There are two types of tattoo kits: a hand-held plier type and a battery-operated pen type.

Before using either type of tool, I always disinfect the ear with a cotton ball dipped in rubbing alcohol. Some breeders find that shaving the inside of the ear with animal clippers beforehand helps apply the tattoo more easily, especially when using a tattoo pen. Some also use a pet-friendly anesthetic spray to dull the sensation in the ear; just be sure to follow the instructions on the packaging. After the procedure, I like to apply a pet-friendly antibiotic ointment to help prevent infection.

The **hand-held plier type** simply requires that you insert the numbers or letters in the pliers and squeeze quickly and firmly on the rabbit's left ear. The needles will puncture the skin. Release the pliers and then rub tattoo ink over the puncture using a disinfected finger or brush. You can let the excess ink dry and flake off or wipe it away with a clean paper towel.

Hand-Held Plier

It's hard to learn the amount of pressure to apply to the pliers to mark the ears. For this reason, practice several times with a piece of cardboard. Some breeders suggest squeezing the plier for two seconds, but I generally find one second is long enough and less painful for the rabbit.

+

- The pliers are quick and efficient.
- The pliers are cheaper than a battery-operated tattoo pen.

✕

- The procedure is painful.
- It is not uncommon for the ear to bleed excessively.
- Sometimes, with younger rabbits, the needles may puncture through the ear.

The **battery-operated pen type** is designed specifically for drawing directly onto the ear for tattoo purposes. Generally, this method is preferred by breeders who plan to exhibit their rabbits in shows, as the ARBA rules call for a "clear, legible tattoo."

Even though this tool is similar to using a pen, it may take some time to get used to writing on the rabbit's soft, pliable skin. For this reason, practice writing with the tool several times with a banana or an orange until you have mastered the technique.

+

- It is simple to use.
- The tattoo is more legible.
- There is less risk of infection.

✕

- It costs almost twice as much as the pliers.
- It takes longer than the pliers.
- It may take extra time to clean and sanitize the tools after use.

HOW TO TATTOO A RABBIT'S EAR

Always work in a clean environment. Often, I will place the rabbit into a carrier, such as for cats, and bring the rabbit onto my porch or inside my house. I recommend you wear a pair of disposable plastic gloves, use disinfectant on the tools between animals, and use new cotton balls on each animal.

Be diligent in cleaning your tools after use to avoid dirt buildup, which could damage the pen or the numbers and letters used in the pliers, resulting in an illegible tattoo.

1. Have all your equipment within arm's reach.

2. Loosely wrap the rabbit in a towel like a burrito, leaving the head exposed.

3. Using a cotton ball, sanitize the inner part of the left ear with rubbing alcohol where the tattoo is to be placed.

4. If you are using an anesthetic spray, apply it now and wait two to three minutes for the ear to numb.

5. If you are using tattoo pliers, be sure that the letters and numbers have been loaded correctly.

6. Place the rabbit in front of you, and, using your writing hand, position the pliers in the center of the left ear and press firmly. Release and lift the pliers straight out to avoid scratching the tattoo. Be sure to follow the instructions on how to use the pen and write the number and letters required.

7. Apply the ink directly onto the tattoo and use a brush or your finger to rub the ink into the puncture marks. Many tattoo pens apply the ink as you write, so this step may be unnecessary.

8. Disinfect all the tools used before proceeding with the next rabbit.

9. The ink will flake off slowly. It will not be harmful to the rabbit if accidentally consumed.

IDENTIFICATION NUMBERS AND LETTERS

There are many systems for selecting ear numbers.

For example, some breeders select the name of the rabbit. You can select short names and tattoo the entire name in the left ear. Three- to four-letter names work the best, such as Sam, Bert, Bugs, and Anna. If your rabbit has a longer name, like Fizzy Pop or Thumper, then you can abbreviate the name.

Personally, I put my initials in the left ear, followed by three to four numbers that indicate the month and year the rabbit was born. For example, if I write DC923, it stands for Candace Darnforth (the breeder), and the rabbit was born in September 2023. Other breeders place the initials F or M to indicate female or male. Or they put the initials of the parents—the buck and the doe. Basically, you can use any system of coding that you find useful.

It is a personal choice if you want to mark your meat rabbits or not. Some breeders tattoo the letters MEAT into the left ear to discourage another breeder from using their rabbits in a breeding program.

Common questions about tattooing your rabbits

What if the tattoo is not clear and legible?

If the tattoo is illegible once dry and the puncture marks have healed, then additional identification marks may be needed. Often, a veterinarian will need to do the second tattoo to avoid the current tattoo so it will not be distorted.

Does tattooing a rabbit's ears hurt?

Tattooing your bunny's ears is similar to having your ears pierced, and it is always best to do it with younger rabbits. A younger kit will have thinner cartilage, meaning it will be easier to tattoo and will heal faster and hurt less.

Why the left ear?

The tattoo number is generally for your identification purposes. The right ear is reserved for an ARBA registration number, which the rabbit may qualify for once it reaches maturity and passes the registrar's examination.

Breeding Methods

As you expand your rabbitry, you will choose a breeding system that will sustain your rabbit production for years to come. There are a few different methods, such as inbreed, linebreed, outcross, and crossbreed.

Inbreeding involves breeding two rabbits that are close relatives, such as a brother and a sister, or a mother and a son, or a son and a cousin, and so on.

Linebreeding is similar to inbreeding as the rabbit is bred to a relative. But it follows a line of descent, most often a grandparent, grandson, aunt, or nephew, and so on. Linebreeding is a more selective method of inbreeding on a line of descent from rabbits of outstanding quality.

Outcrossing involves breeding the rabbit with an unrelated animal but of the same breed. This method of breeding is preferred for new rabbit farmers. However, most breeders will outcross their rabbits from time to time, especially if they plan on selling breeding stock or showing their rabbits.

Crossbreeding involves breeding two different breeds. Many successful breeders of meat rabbits prefer this method as it can produce a higher meat yield. However, if you want to improve existing breeds or establish a new variety, this method is best left to experts.

WHY DO SOME RABBIT FARMERS OPT FOR INBREEDING?

For humans, the thought of inbreeding—the mating of close family relatives—is taboo, and inbred human offspring are riddled with a long list of health issues.

So why do meat rabbit farmers inbreed? Inbreeding accentuates existing characteristics, both good and bad. For example, mating two rabbits with desirable traits may enhance those qualities in the offspring, though this is not guaranteed. If the offspring don't show improvements, inbreeding can still help reveal problem areas in your line that may have been previously masked. On the other hand, breeding two rabbits with undesirable traits often intensifies those flaws, so careful selection is essential.

If you decide to cross two closely related rabbits, first, you need to carefully select the doe and buck that have the characteristics that you desire in offspring.

For obvious reasons, inbreeding brings no new blood into your rabbit colony. If your foundation stock is of poor quality, then you are better off introducing new blood. Generally, I will start off with outcrossing by using superior foundation stock, then limit the gene pool to maintain that quality of line in future generations.

If you plan on crossing two pairs of distantly related rabbits, the two pairs will result in two different litters. You will want to save three does and two bucks from these litters. You want to increase the size of your breeding stock, so you will breed the two pairs again, and once again, you will save three does and two bucks from these litters. Now, when these rabbits reach breeding age, you may:

- Cross a buck from one litter with a doe from another
- Cross the doe from the first litter with her sire
- Cross a doe from the second litter to the other litter's sire
- Cross the buck back to its dam
- Cross the buck to the other litter's dam

These possibilities can yield more litters in less time, meaning more meat rabbits to be processed and sold. Inbreeding your colony is a sustainable way to produce rabbits at a low cost for a long period of time. However, I do not recommend inbreeding for a long period of time, as the bloodline can become weaker. For this reason, I introduce new blood into the herd every year or two.

SELECTING FUTURE BREEDERS

Let's say you produced nine litters over a period of several months by inbreeding your colony. Any of the above scenarios would produce dozens of potential candidates for future breeding stock. But how can you select the best rabbits for future breeding?

Each prospective breeder should be of equal quality, preferably exceeding the selection qualities of its dam and sire. The rabbits that do not measure up to your requirements can be processed and sold off for profit. Never mate two rabbits who have the same weaknesses, as the litter will have a double dose of that same issue. On the other hand, if you carefully select your breeders, the litters will surpass your expectations.

Each rabbit farmer has their own standard for breeding stock. With time, you will recognize what is better and ingrain it in your colony. When judging whether a rabbit is good breeding stock or not, take into consideration your breeding goals, the breed standard, and the rabbit's overall health.

CHAPTER 9

Preventative Medical Care

U nfortunately, just because you have top-quality breeding stock does not mean your rabbits will be immune to diseases and other health issues.

Disease Prevention

Always take proactive steps to prevent health issues because a recovered rabbit seldom returns to full health. In this chapter, we will be discussing how to stop problems before they start. Here are some suggestions to keep your rabbit herd thriving.

Besides feeding your rabbits good-quality pellets, it's important to choose a protein level appropriate for their stage of growth and purpose. Growing kits and lactating does typically require a higher protein content, around 16% to 18%, to support development and milk production. Maintenance diets for adult rabbits not in breeding or production generally contain 12% to 14% protein.

While some rabbit farmers feed a higher-protein diet (16% to 21%) to accelerate weight gain, excessively high protein levels can reduce fiber intake, potentially leading to digestive issues. Overfeeding protein can also contribute to excessive fat, which may impact reproductive health in breeding stock. I prefer feeding rabbits a balanced, high-fiber diet with moderate protein (14% to 16%) to promote steady growth and lean muscle development.

If your weaned young stock gets diarrhea, it is often is fatal, and if not, they rarely develop properly, so you would be better off culling the rabbits.

Another method to prevent diseases is to ensure that your pellets contain copper sulfate. Copper sulfate helps to prevent intestinal issues; we will talk more about this later in this chapter. Here are some other methods to prevent illness or diseases.

FEEDING

Keep a feeding schedule. Never feed fresh greens to younglings because it can make them very sick or even be fatal. A young rabbit's stomach cannot handle abrupt changes; for example, say it eats a regular diet of pellets, hay, and water, and then you give it something wet and grassy. You will end up with a very sick, or more likely dead, rabbit.

WATER

Be sure to give your rabbits fresh, clean water daily. As mentioned previously, rinse out the crocks daily and disinfect them weekly. If you have an automatic watering system, check the valves daily and flush the system once a month or more if you have hard water, as it can cause a buildup of sediment. Rabbits will not eat if they are not getting enough water.

ISOLATION

Even if you are bringing home new healthy stock with bright eyes, glossy coats, firm flesh, and vigor, you will want to isolate them. Place the rabbits in a pen away from your rabbitry, such as inside your garage or tool shed.

OBSERVE DROPPINGS

Check your rabbits' droppings daily. Healthy droppings should be round, firm, and large. Take immediate action at the sign of diarrhea. Change the bedding, water, and pellets in the sick rabbit's cage and isolate the rabbit from the rest of the herd.

HEALTH CHECK

Weekly, check each rabbit for bright eyes, a glossy coat, firm and pliable skin, and active behavior. Examine footpads for signs of sore hocks or ulcers; starting with a resting pad in the hutch can help prevent these issues before they appear. Also, watch for any runny noses or sneezing. A rabbit with these symptoms will often have wet, matted fur on its front paws from wiping its nose.

DO NOT BE AFRAID TO CULL THE RABBIT

If you have a rabbit with a cold, pneumonia, or any other illness, do not hesitate to cull the animal. Often, it will cost you a small fortune to cure, and diseased rabbits can infect the entire herd. Most diseases can be prevented, but others can knock out an entire herd if you do not act quickly.

Common Health Issues

While rabbits can be relatively hardy, they are prone to certain health issues. Common problems include dental issues, gastrointestinal issues, respiratory infections, and parasites.

Providing a well-balanced diet and a clean environment can help prevent many illnesses in rabbits. Prompt attention to any signs of sickness is essential for their well-being.

HEALTHY VITAL SIGNS FOR RABBITS

According to the *Merck Veterinary Manual*, these are normal vital signs for a rabbit:

Body temperature:	101.5–103°F (38.6–39.4°C)
Rectal temperature:	103.3–104°F (39.6–40°C)
Heart rate (pulse):	130–325 beats per minute
Respiratory rate:	32–60 breaths per minute

GASTROINTESTINAL ISSUES

The most common health issue in rabbits is gastrointestinal distress, such as diarrhea. It may be highly contagious and can wipe out your entire herd in less than a week.

Coccidiosis is caused by a parasite called coccidia. The parasite prevents the rabbit from converting food properly, so the animal is unable to gain weight. This condition leaves spots on the liver, making the processed rabbit unsuitable for sale. This parasite weakens the rabbit's immune system, making it more susceptible to other health issues.

Mucoid enteritis causes severe diarrhea in rabbits ages five to eight weeks. This condition can kill off a healthy rabbit in less than 24 hours. One night, you have a healthy bunny bouncing around its hutch. The next morning, the rabbit is weak, thin, and scruffy-looking, wet around the mouth, sitting in the corner listlessly. The rabbit emits a constant stream of foul-smelling feces with a jelly-like consistency. By evening, the rabbit will have passed away. And the following day, many other rabbits will follow.

You can lose the entire litter if you do not act fast in culling the sick rabbits and isolating the healthy ones.

Copper sulfate is essential in preventing both the aforementioned conditions. Besides feeding your younglings a diet of low-protein, high-fiber pellets, be sure to mix in some easily digestible oats and low-protein hay, such as Timothy.

As mentioned in previous chapters, before a litter is born, the hutch and nest box should be thoroughly disinfected.

RESPIRATORY ISSUES

If you notice any of your rabbits sneezing, with mucus dripping from the nose or front paws that are wet and matted, then you will need to quarantine the rabbits to separate living quarters from the rest of the herd, such as in your garage.

If the relocated rabbit stops sneezing, perhaps the rabbit had an allergy to moldy hay or dust from the rabbit pellets. Or maybe the rabbit was sneezing because of poor ventilation and a high ammonia level due to a buildup of urine. You can improve the ventilation, remove the moldy hay, and use self-sifting, screen-bottom feeders.

If the relocated rabbit slowly improves, it may have had just a cold or the snuffles. There are no reliable treatments or ointments for a cold or the snuffles, as they just mask the symptoms. Your best option is to keep the rabbit isolated for two or three weeks after the symptoms. If the symptoms stop, then it should be safe to return the animal to the herd.

On the other hand, if the symptoms persist, you will need to cull the animal and thoroughly disinfect its hutch and feed and water dishes. Otherwise, you may end up infecting the entire herd and having to cull all the animals.

The United States Department of Agriculture recently approved a vaccine to prevent the Pasteurella infection that causes the snuffles.

For more information, you can visit www.pavlab.com

OTHER ISSUES

Sore hocks

Rabbits can develop sore hocks, also known as pododermatitis, due to factors like improper flooring, lack of exercise, or obesity. It's important to provide a clean and padded living environment to prevent this condition.

If your rabbit develops sore hocks, treat them using a tincture of iodine or an antibiotic ointment. Place a wooden board inside the rabbit's hutch for it to sit on. Often, rabbits that develop sore hocks have thinly furred footpads and are not suitable candidates for breeding as they may pass this weakness on to their offspring. Also, rabbits who tend to stomp their back paws frequently suffer from sore hocks. Once again, I would not recommend breeding this characteristic into the herd.

Ear Cankers and Fur Mites

Ear cankers are scabs on a rabbit's ears caused by parasitic mites, which thrive in dirt and manure. Rabbits housed on solid dirt floors are more susceptible, while those in elevated hutches with wire bottoms are less likely to be affected. Fur mites, another common parasite, can cause itching, hair loss, and skin irritation along the rabbit's body, especially around the neck and back. Both ear and fur mites can be treated by applying mineral or vegetable oil; drop a few drops into the ear for ear mites or rub a small amount along the affected area for fur mites. The oil suffocates the mites, leading to relief and healing within a few days as the crusty scales and irritated skin begin to recover.

Ringworm

Ringworm is the only rabbit ailment that is contagious to humans. This fungal infection is caused by various species of fungi known as dermatophytes. Despite its name, ringworm doesn't involve worms but rather creates red, ring-like rashes on the skin. It can affect rabbits, leading to skin irritation, hair loss, and itching.

Treatment usually involves antifungal medications, both topical and oral, prescribed by a healthcare professional or veterinarian. Regular cleaning of the rabbit's living area can help prevent the spread of the infection.

Dental issues

Rabbits commonly face dental issues like overgrown teeth, malocclusion (misalignment), and dental spurs. These problems can result from a diet lacking in fiber, causing inadequate wear on their continuously growing teeth. Regularly providing hay and appropriate chew toys can help promote dental health in rabbits.

If you notice signs like difficulty eating or drooling, the rabbit should be culled, as it will not be able to eat enough to reach an adequate weight for processing. Also, I do not recommend using a rabbit with misaligned teeth for breeding, as the offspring will most likely inherit this flaw.

Sanitation

Healthy rabbits are the key to successful rabbitry. However, your success depends on maintaining clean, dry hutches in a well-ventilated rabbitry.

All-wire hutches are designed to keep the inside of the hutch clean and dry as the urine and manure drop through the wire flooring onto the ground. The urine and manure land in a pit with gravel, sand, or small pebbles that drain away the urine, keeping the manure dry and preventing a swarm of flies.

As the manure decomposes, worms will break it down into rich, potting soil, which reduces the odor. Typically, I only remove the manure once a month. I move it into a compost pile so the worms can do their job. Once a week, during the warmer months, I sprinkle lime or superphosphate on the manure to reduce odor and help improve the quality of manure to be used as fertilizer.

A damp, smelly rabbitry is a recipe for disaster. Dampness can lead to various health issues for rabbits because it promotes the growth of mold, bacteria, and parasites. This can cause respiratory problems, skin infections, and other illnesses in rabbits. Additionally, dampness can make the living space uncomfortable and contribute to stress, negatively impacting the rabbits' overall well-being. Furthermore, flies breed in moist manure, and flies transport a long list of illnesses that can make your herd sick.

Typically, it is normal to have a fly or two in the rabbitry. But during the warmer months, flies might multiply. To keep the flies at bay, you will need to spray the manure beds down with a fly spray. When spraying the manure, take caution to avoid spraying the rabbits' feed and water. Typically, I remove the water crocks and empty the feeding dishes to avoid contamination.

Abscesses

An abscess is a bacterial infection that may be caused by a number of reasons, such as dental disease or injury. Abscesses are easy to diagnose as the skin will be swollen, containing an accumulation of pus. Abscesses in rabbits are common under the skin around the jaw or wherever there is a scratch or an open wound.

If you have ever had an abscess, then you already have a general idea of how to treat the wound, such as antibiotic ointment and lancing. Lancing requires opening the wound with a disinfected needle or blade and removing the pus. Once the pus has been drained, the wound should be disinfected with peroxide and packed with antibiotic ointment.

In most cases, I cull a rabbit with an abscess as it is an indication of a weakened immune system.

Blocked mammary glands and mastitis

Occasionally, nursing does produce more milk than the litter is able to drink. This causes a blocked mammary gland, or in layman's terms, a blocked duct.

Examine the doe for swollen, hot, and firm teats. Blocked mammary glands are easy to remedy; simply massage the gland with a warm, humid cloth to reduce the buildup of milk by partially milking. Make sure not to drain all the milk, as that will stimulate more milk production.

Mastitis is a bacterial infection that is often confused with a blocked mammary gland as the teats appear red and swollen and are hot to touch. The colloquial term for mastitis is blue breast, as the tissue around the teat appears to have a blue tone from bruising. This condition can be very painful for the doe, and she may refuse to nurse her litter.

Antibiotics are needed to treat this condition. For this reason, it is essential to prevent the condition by making sure the nest box does not have any sharp edges that may scratch or cut the doe's teats. Be sure to sanitize boxes between litters, even if the same doe is using the box. Do not move kits from the infected doe to a foster doe, as this may spread the infection. Do not forget to disinfect your hands or tools before handling other rabbits.

Rabbit syphilis

Rabbits, like people, can get venereal diseases, such as syphilis. Watch your rabbits for blisters, scabs, and inflammation of the genitals. Also, red sores or blisters around the mouth may be an indicator. Both sexes can be infected by this condition.

Breeding does may pass this condition onto their litters, so you will need to get the outbreak under control quickly. In order to treat your rabbits, you will need to get an appropriate antibiotic from your veterinarian. All rabbits that have been in close physical contact with one another will need to be treated, even if they are not displaying symptoms. Culling just the sick animal is often not an option, as the majority of the herd will have been infected because their hutches are side by side.

Hutch burn

Also known as urine burn, this is caused by constant contact with urine. Hutch burn is common in rabbits that live in solid-floor hutches as they cannot be cleaned properly; however, dirty resting pads or nesting boxes may also be the culprit. This condition is easy to identify as the rabbit will have bald patches with red, inflamed skin, especially around the hind legs and genital region.

Since hutch burn is caused by environmental factors, it can easily be remedied by cleaning the hutch and ensuring the floor stays dry and clean. To treat hutch burn, rinse the affected rabbits with warm, clean water and pat dry before returning to their clean, dry hutch.

Nutritional Deficiencies

Nutritional deficiencies in meat rabbits can lead to various health problems. Common issues include the following:

- Vitamin Deficiencies — Lack of essential vitamins like vitamins A, D, or E can affect overall health, growth, and bone development.
- Mineral Deficiencies — Insufficient minerals, such as calcium or phosphorus, can impact bone strength and development.
- Protein Deficiency — Inadequate protein intake can hinder growth, reproduction, and overall muscle development.

Ensuring a well-balanced diet with quality hay, fresh vegetables, and appropriate commercial rabbit pellets is crucial to prevent nutritional deficiencies in meat rabbits. Regular veterinary guidance can help tailor the diet to meet specific nutritional needs.

Hereditary diseases

Hereditary diseases in meat rabbits can vary, but some potential examples include dental issues, such as malocclusion (misalignment of teeth), which may impact their ability to eat. Additionally, conditions like certain respiratory or cardiac disorders may have a genetic component. Responsible breeding practices can help identify and manage hereditary issues in meat rabbit populations.

WHAT TO DO WHEN YOU DO NOT KNOW WHAT TO DO

If you cannot diagnose what is wrong with your rabbits, contact a vet. As we will discuss below, not all vets specialize in rabbits and often suggest a course of treatment that does not make sense for a small rabbitry because it is too expensive or requires the use of medicine that is not suitable for meat processing.

Another option is to reach out to the breeders of your foundation stock or other experts in raising meat rabbits to see if they have encountered anything similar and ask how they treated the condition. Also, there is an extensive online community of experts and specialists, such as the livestock experts at the Cornell Cooperative Extension, that you can reach out to for advice.

If you are still stumped, you can bring a deceased rabbit to a diagnostic laboratory in order to discover the exact cause of death. The American Association of Veterinary Laboratory Diagnosticians has a comprehensive list of labs organized by state (www.aavld.org/accredited-laboratories). However, the service is not free! On average, it costs about $250 per specimen to be diagnosed.

In most cases, a livestock diagnostic lab is not close by, so you will need to mail the frozen carcass of the animal. Your local veterinarian will be able to provide you with the required information for shipping a frozen carcass of a diseased animal via courier.

Be sure to provide the pathologist with as much information regarding the deceased rabbit as possible, such as the number of rabbits in the rabbitry, number of sick or deceased rabbits, age and sex of the affected rabbits, description of the disease as it progressed day by day, and the dates of first and subsequent losses. Also, you can include details of what type of treatment was applied, infection incidence, and any other relevant information that may help explain the outbreak.

It will take a week to receive the results of the test. There is no need to send every deceased specimen to a diagnostic lab. Often, this is only done if you suspect you have an unknown outbreak in your rabbitry. Use your discretion.

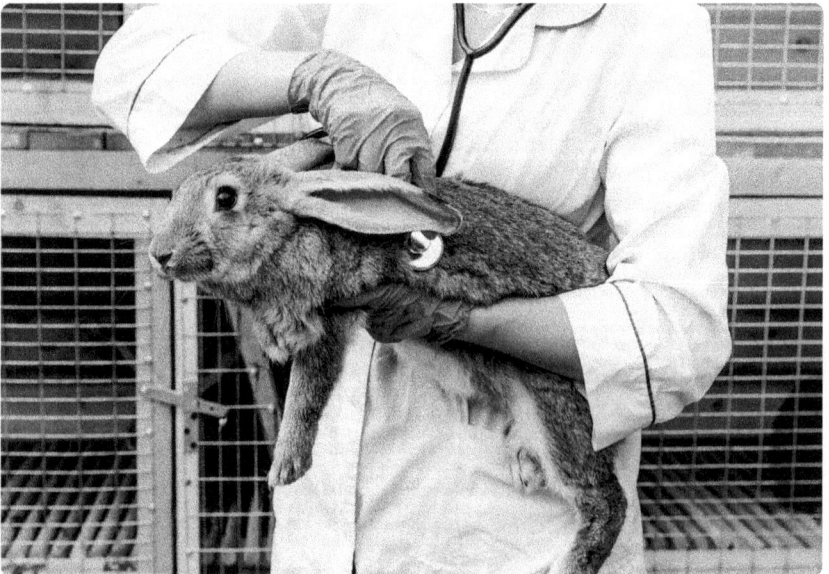

Medical Care

Generally speaking, most rabbit farmers avoid using veterinarian services due to high costs, limited access to veterinary care specializing in rabbits, and the fact that rabbits are hardy animals, requiring minimal medical attention. Most of the time, a sick rabbit will be swiftly culled.

If, for whatever reason, you need professional medical care for your rabbits, choose a veterinarian carefully. Typically, veterinarians examine rabbits using techniques similar to those used with cats and dogs. Routine vaccinations are not currently required for rabbits in the US.

Signs of illness may include one or more of the following:

- Discharge from the nose and eyes
- Fur loss
- Red or swollen skin
- Loss of energy, weight, or appetite
- Drooling
- Diarrhea or no droppings for more than 12 hours
- Disorientation (not hopping or moving normally)
- Trouble breathing
- Chattering or grinding teeth while sitting in a hunched position

If your rabbits show any of these symptoms, you should quarantine them away from the rest of your herd. If the symptoms do not improve, you may need to eliminate the afflicted animals.

There are very few drugs approved for use on rabbits. For this reason, veterinarians tend to use drugs approved for other species of a similar size, such as felines and canines. Extreme caution should be used when administering antibiotics. Some antibiotics can cause an imbalance in intestinal bacteria, severe diarrhea, or even death. The following antibiotics should not be used

in rabbits: cephalosporins, amoxicillin, clavulanic acid, lincomycin, erythromycin, ampicillin, and clindamycin.

Flea treatments should also be avoided as the common ingredient, fipronil, is fatal for rabbits if consumed.

Prolonged fasting prior to a surgical procedure is not required or recommended. Rabbits are unable to vomit. Vomiting is a general concern for other species during general anesthesia. It is essential the rabbit eat immediately after surgery. Hay and water should be offered as soon as possible after surgery. Bananas or alfalfa hay may increase the rabbit's appetite after a surgical procedure.

PROTOCOL FOR A SICK RABBIT

No matter what the issue is, there are six steps you should take upon encountering a sick rabbit.

1. Mark the hutch that contains the sick rabbit for cleaning and disinfecting later.

2. Place the sick rabbit in isolation. Ideally, move the rabbit into another building, but if that is not possible, place the rabbit as far away as possible from the rest of the herd. There should be no direct contact between animals.

3. Before handling healthy rabbits, disinfect your hands, boots, or tools. Disinfection is your best defense to prevent the spread of disease.

4. Diagnose the problem and begin the appropriate treatment. If you are unable to make a diagnosis, contact a veterinarian.

5. Cull any sick rabbits that have no hope of recuperating. Bury or burn the animal carcass away from your rabbitry and farm.

6. Clean and disinfect any hutches and watering and feeding dishes the sick rabbit came into contact with.

> "Unfortunately, it's hard to find a vet that will treat rabbits, let alone treat rabbits as livestock... When you're looking for a vet to work with, it's extremely important for the vet to know they are a meat rabbit, not a pet, and your goal."

Natasha Spudville, D'Argent Rose Rabbitry

For most rabbit farmers, using a vet regularly can be cost prohibitive. While it's always good to have a vet in mind for emergencies, many farmers choose to handle routine care and minor medical issues themselves. If a rabbit becomes seriously ill and home care isn't effective, it may be more practical to cull the rabbit rather than spend significant money on treatment, especially if the rabbit is valued primarily for meat or general breeding purposes. However, in rare cases—such as with a valuable prize buck or doe—seeking veterinary care may be worth the cost. Here are some considerations if you decide to work with a vet.

Seek Word-of-Mouth Advice

Ask fellow rabbit farmers or breeders in your area for recommendations. Other local farmers can often suggest vets who are familiar with rabbits, understand common rabbit health issues, and offer practical advice.

Find a Vet with Experience in Treating Rabbits

Not all veterinarians have experience with rabbits, so it's important to find someone familiar with their specific needs. While it may not be necessary for every visit, having a vet with rabbit expertise as an occasional resource can be helpful.

Consider Cost and Location

In the rare instance you may need urgent care, choosing a vet within a reasonable distance is ideal. Additionally, given that routine vet visits are uncommon for many rabbit farmers, it's wise to inquire about costs upfront to ensure they align with your budget.

Focus on Practicality Over Routine Visits

Since regular vet visits may not be financially viable, consider using a vet primarily for advice on preventive measures or in extraordinary situations. Take note of the clinic's cleanliness and professionalism, as well as any specialized rabbit knowledge they offer.

Personal Referrals and Basic Research

Even with a trusted referral, take some time to research the vet's qualifications and experience. Building a relationship with a veterinarian who understands the practical needs of a rabbit farm can be valuable if unique situations arise.

In most cases, the best approach is to focus on good preventive care and health monitoring in your own rabbitry, relying on a vet only for occasional support or emergencies with valuable rabbits.

> *"I refer to The Rabbit-Raising Problem Solver by Karen Patry for general health issues or breeding dilemmas. For anything that would require veterinarian attention, we typically opt to just cull the rabbit. Communicable diseases can be spread easily in a large herd, so sacrificing one sick rabbit is often preferable to possibly having all the rabbits come down with the illness."*
>
> **Toni Case**, Lost Mountain Farm

Human Foods Unsafe for Rabbits

Rabbits are generally not picky eaters and will munch on whatever you put in their hutch. But some foods can cause serious health issues or even be fatal. Most of these foods are obvious, but you might be surprised at some of the foods your furry bunnies need to avoid at all costs!

Alcohol — Even the tiniest amount of any type of alcohol can be fatal for rabbits. Alcohol causes rabbits to have coordination problems, vomiting, diarrhea, breathing issues, and even death.

Avocado — All rabbits are allergic to persin, which is found in high quantities in avocados. Persin is not only found in the flesh of the avocado but also in the leaves, peel, bark of the tree, seed, etc.

Chocolate — Dark, white, and milk chocolate is deadly for rabbits. Even the smallest morsel can cause diarrhea, vomiting, cardiac failure, seizures, and even death.

Dairy products — Dairy products such as milk, cheese, whipped cream, and ice cream can cause your rabbit to experience digestive discomfort and diarrhea.

Garlic and onions — Keep all types of garlic and onions, including fresh, dry, powdered, dehydrated, or cooked, far away from your rabbit. Even the smallest pinch can cause your rabbit's blood count to drop, causing it to become anemic.

Kale — Kale contains large amounts of oxalates and goitrogens, which may cause bladder sludge and other serious health issues.

Grapes or raisins — Grapes and raisins may seem like the perfect bite-sized treat for your rabbit, but just a few can cause kidney failure. If you think your rabbit may have consumed some grapes or raisins, call your veterinarian if you notice any sluggish behavior or severe vomiting.

Macadamia nuts — Eating just one macadamia nut can cause your rabbit to become seriously ill. Eating chocolate-covered macadamia nuts will intensify the symptoms, which will eventually lead to death. Macadamia nuts cause vomiting, muscle tremors, fever, and loss of muscle control.

Pitted fruits — Fruits such as peaches, persimmons, cherries, and plums have pits or seeds that can get lodged in your rabbit's intestines, causing a blockage. Some pits, such as from a plum or a peach, contain cyanide, which is fatal if consumed by a rabbit.

Salt — A word of caution: do not share your salted popcorn or pretzels with your furry friend.

Too much salt can cause sodium ion poisoning, vomiting, diarrhea, fever, or seizures and may be fatal if left untreated.

If your rabbit ate something it shouldn't have, call your local vet immediately or call the Animal Poison Control Center (ASPCA) at (888) 426-4435.

It is not advisable to make a practice of giving your rabbits leftovers, bits of meat, or other scraps, as research shows these items may contribute to fatal cases of enterotoxaemia, a toxic overgrowth of bad bacteria in the digestive tract.

ASK THE EXPERTS

Have you ever experienced significant disease or health issues in your rabbits? If so can you explain what happened and how you handled it?

Health issues are an inevitable part of raising rabbits, and our experts shared their experiences and strategies for managing common diseases and maintaining a healthy herd. From dealing with ear mites and GI stasis to taking biosecurity measures and making tough culling decisions; managing rabbit health requires vigilance and care. Below, we summarize the key health challenges breeders face, along with practical advice on prevention, treatment, and the emotional aspects of maintaining a healthy rabbitry.

1. COMMON HEALTH ISSUES

Common health problems encountered by breeders include ear mites, coccidiosis, malocclusion, and gastrointestinal (GI) stasis. Ear mites can be treated effectively using topical ivermectin or other mite treatments, while coccidiosis often requires targeted treatment with medication like Corid or Toltrazuril. GI stasis, which can be fatal, requires rapid intervention to restore gut movement.

"The most common health issues seen in rabbits are malocclusion, digestive upset, pasteurella, ear mites, and coccidiosis. Pasteurella is a disease that cannot be cured, but is present in a significant percentage of rabbits... Any animal suspected of pasteurellosis should be immediately quarantined or culled."

Heather Riddell-Ide
Riddell-Ide Farm

2. BIOSECURITY MEASURES

Maintaining good biosecurity is essential for preventing the spread of disease. Quarantining new rabbits for a period of 2-4 weeks helps detect and prevent the introduction of illnesses into the main herd. Limiting access to the rabbitry and ensuring visitors wear clean clothing can also help in maintaining a healthy environment.

"When we bring in new stock, we are very careful to have them in a separate area for 2-4 weeks so we can observe them for any unknown illnesses, and so they don't spread that to our other rabbits. We will take care of our resident rabbits first before we care for any new arrivals."

Melina Anderson
Shining Light Farm

3. CULL DECISIONS AND PREVENTION

In some cases, breeders have had to cull rabbits due to illness or genetic issues. Conditions like pasteurella or malocclusion are often best addressed through culling to prevent the spread or perpetuation of these problems within the herd. Preventative measures, such as keeping the rabbitry clean and well-ventilated, help reduce the overall risk of disease.

4. WEANING CHALLENGES AND COCCIDIOSIS

The weaning stage can be a critical period for young rabbits, with digestive issues such as bloat and coccidiosis being common. Providing hay during weaning helps support gut health and prevents bloat. Maintaining cleanliness is key to minimizing the risk of coccidiosis, which can be particularly harmful to young kits.

"Bloat can be a big problem for kits during the weaning stage if hay isn't offered. We have had an issue with white nasal drainage and learned that is a cull issue. We tried everything but nothing made it better. Cull the animal no matter how good the genetics are and move on."

Jason Lightfoot
Deer Run Rabbitry

5. EMOTIONAL ASPECTS OF CULLING

Culling is often one of the most emotionally challenging aspects of rabbit farming, especially when it involves a favorite breeder or an emergency situation. Many breeders mentioned the difficulty of making culling decisions but emphasized that it is necessary for the health of the herd. Proper preparation and a humane approach are crucial during these difficult moments.

CHAPTER 10

Processing Your Rabbits

In this chapter, we delve into the intricate art of processing meat rabbits—from humane handling to precise butchering. Prepare to embark on a journey that not only respects the animals but also provides a deeper connection to the food on our plates.

The Unspoken Reality

If rabbits happen to be your first experience in raising animals for meat, coming to terms with killing an animal you have raised can be emotional. It is hard to not feel a twinge of sadness, but it definitely becomes easier over time.

I recall the first time I slaughtered my rabbits; I was conflicted. On one hand, I was excited to butcher a rabbit all by myself for the first time. On the other, I felt sorry to see my furry little bunny go. Over the years, I have come to the conclusion that conflicting emotions are nothing to be ashamed of.

To be honest, all the meat we have eaten has had eyes, a heart, and a unique personality at some point—whether you cut it up and put it in the freezer or someone else does. Unfortunately, we have become detached from this concept. To the majority of people, hamburger comes from hamburger. It is most definitely not from a living, breathing cow. We have been programmed not to think about that part.

For first-time rabbit farmers who are anxious about their first time butchering a rabbit, here are a few suggestions that help silence the moral quandaries of killing an animal for food.

1. Homestead rabbits generally live a good life, especially when compared to their industrial counterparts. For example, my rabbits have oversized cages that are cleaned daily, plus I give each rabbit at least 20 minutes each day in an enclosed pasture to forage, hop, and play.

2. Home-raised rabbit meat is fed higher-quality feed, meaning the animals are healthier and happier. Our rabbits are fed the best-quality pellets available, plus they have a constant supply of hay to munch on and a daily treat of fresh greens. My rabbits never eat any by-products like most commercial rabbits do. That equals a healthier rabbit and better-quality meat for us.

3. Farmed rabbits have more peaceful deaths. My rabbits breathe their last breath only about 20 feet from their hutch. Their death is swift and calm. Rabbits live in the moment and are incapable of knowing what is coming next.

4. We need to get our meat from somewhere. Why not from a sustainable rabbit farm that raises happy bunnies? Without a doubt, I would rather eat an animal that I know was fed properly and treated humanely instead of "mystery meat" from the grocery store.

There is no need to be wary of your first time butchering your rabbits alone. There is no reason to be embarrassed about feeling a little sad when you butcher Mr. Bunny. Acknowledge the feelings, enjoy the experience, and take pride in learning and perfecting a new skill that is fast becoming extinct in our modern society.

Introduction to Butchering

Most meat eaters state they could never slaughter an animal. They most likely feel this way because they do not know how to do it.

Honestly, I can remember how awkward I was the first time I butchered an animal. I must have stood with the knife in my hand for at least 30 minutes before I could even attempt to use it. Just the thought of slaughtering the rabbit terrified me, even though a friend had already taught me on two occasions. But this time was different; I was alone with nobody to guide me.

I cannot lie—my first time was neither as quick, smooth, or painless as I wanted it to be. Without a doubt, my lack of inexperience caused the rabbit to suffer a bit more than needed. However, my second time was smoother, my third was better, and my fourth was perfect. Practice makes perfect.

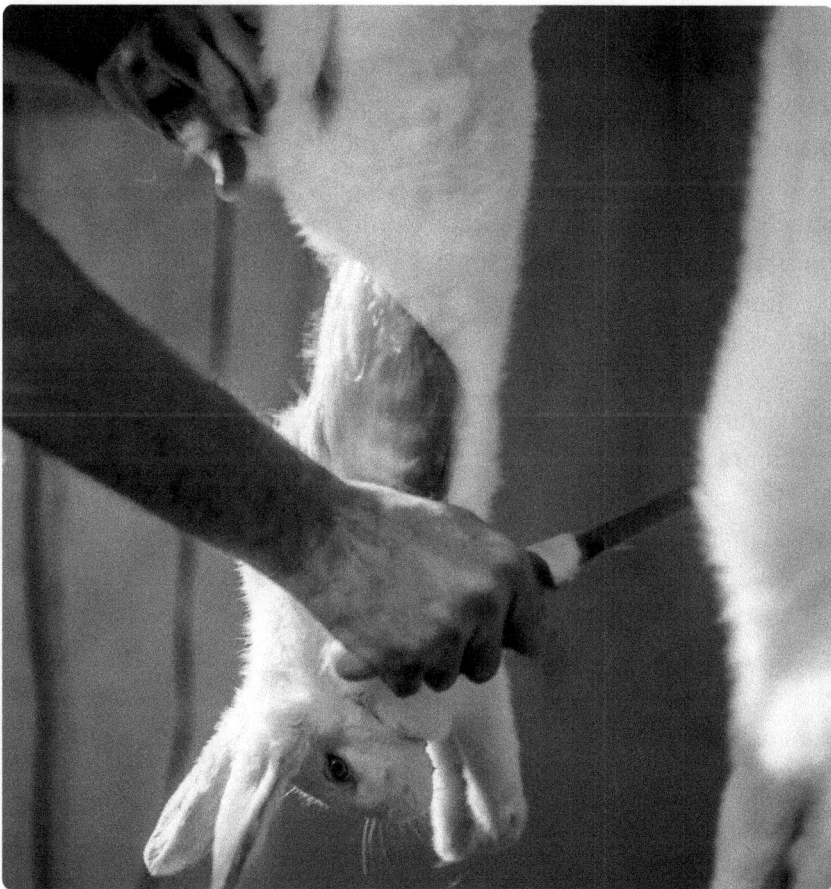

If everything goes according to your calculations, your litter should be ready to be processed at 12 to 16 weeks of age. Typically, during this time frame, I will weigh each individual rabbit and pull those that have reached a live weight of 5.2 to 6.5 pounds. Most often, a rabbit will dress out at 55% of its live weight, so the dressed weight will give me 2.9 to 3.5 pounds for the market for fryers. If your market is for roasters or stewers, you can hold your rabbits a little longer.

The correlation between the weight of a dressed carcass and the weight of the live animal is referred to as the dressing percentage and typically varies for each breed. A dressing percentage of 55% is a standard average by industry standards. However, a herd that exhibits superior meat characteristics can easily achieve a higher percentage. With good genetics and selective breeding techniques, it is possible to get a percentage up to the mid-60s.

With time, you will be able to eyeball the weight of your rabbits, but in the beginning, it is best to get out your scale and weigh each individual rabbit. Let's say you are getting $9 a pound. If you pull too many underweight rabbits, without a doubt, you will be losing money in the long run. Typically, I use a hanging scale with a sturdy box attached that I can place the rabbits into, as weighing a wiggly bunny on a tabletop scale is almost impossible.

Commercial rabbit farms can raise rabbits to market weight in as little as eight weeks through carefully controlled feeding programs and selective breeding for fast growth. In contrast, smaller, sustainable rabbit farms often prioritize slower growth rates, higher-quality feed, and more space for movement. While this approach may not be as efficient as large-scale operations, it emphasizes humane treatment, which many believe contributes to better meat quality, including improved texture and flavor.

Dressing or Processing Your Rabbits

The dressing procedure is also called processing because the process has three main steps. Here is a quick overview of each step in the dressing procedure. Then, we will go into more detail for each step.

Step one — Slaughtering the animal.

Preparing the rabbits for slaughter requires careful planning to ensure a humane and efficient process. By prioritizing the rabbit's well-being, the quality of the meat will be preserved, and the process will be carried out with respect for the animal. When the rabbit has been slaughtered, it is often referred to as being dispatched.

Step two — Draining the blood and skinning the animal.

This step is performed almost simultaneously after the rabbit is dispatched to prevent the blood from coagulating. The skinning process involves carefully removing the hide from the carcass.

Step three — Eviscerating the animal (removing the rabbit's internal organs).

This step is crucial for preparing the meat for human consumption. The internal organs must be removed with care to prevent contamination of the meat. Care should be taken to avoid damaging the meat.

Now, we'll go into more depth about each step.

Due to rabbits' sensitive nature, humane dispatch methods are crucial to minimize stress and discomfort during this process. There are several ways to slaughter a rabbit, but we will focus on the most humane and commonly used methods—cervical dislocation and a fatal blow.

Here are the most common methods to slaughter and dress a rabbit:

1. Broomstick/Rebar Method (Cervical Dislocation)
This is one of the most preferred methods because it is quick, effective, and requires minimal equipment. It instantly dispatches the rabbit, providing a painless and humane end.

1. Place your rabbit on the ground between your feet.

2. Place a sturdy bar, such as a broomstick or rebar, just behind the rabbit's skull.

3. Standing lightly on the bar with your feet, bend over and pick up the rabbit by its hind legs.

4. Swiftly step down on the bar and firmly pull the rabbit's hind legs upward. This movement instantly separates the vertebrae in the neck, killing the rabbit painlessly.

5. Follow the detailed instructions below on how to remove the blood and the hide.

2. Cervical Dislocation by Hand or Choke Chain
Another approach to cervical dislocation involves manually breaking the rabbit's neck by hand or using a choke chain for better leverage. This method is useful for those who may not have the physical strength needed for manual dislocation alone. The goal is to achieve a quick break with minimal suffering.

3. Hopper Popper

The Hopper Popper is a commercial device designed specifically for cervical dislocation. It provides consistent positioning and force, helping to reduce human error and ensuring a more humane dispatch. This method is especially favored by beginners who want a more controlled and repeatable process.

4. Pellet Gun or .22 Rifle

Using a pellet gun or .22 rifle involves delivering a shot to the rabbit's head, usually behind the ears or between the eyes, ensuring an immediate and humane death. This method is effective but requires a suitable environment, such as a quiet area away from neighbors, and proper knowledge of firearm safety.

5. Captive Bolt Gun

A captive bolt gun delivers a strong, focused force to the head, causing immediate unconsciousness and death. This method is similar to those used in larger livestock processing and ensures a quick, humane end.

Note that with all methods of slaughter, even though the animal will be dead, it may still twitch due to neurons firing. The rabbit's eyes may remain open, but this does not indicate awareness—it simply means there are no signals to tell the rabbit to close its eyes. It is also important to handle the rabbit with care before slaughter to minimize stress and ensure the quality of the meat.

After the rabbit is dead, the faster you proceed to draining the blood and removing the hide, the better, as this will prevent the blood from coagulating.

STEP TWO — DRAINING THE BLOOD
AND SKINNING THE ANIMAL

1. Hang the dead rabbit with its skull facing down. Secure it by its hind legs.

2. Using a sharp knife, slit the throat and let the blood drain into a bucket on the ground.

3. Using a sharp knife (about the size of a paring knife), score the skin around the hind leg joints. Using your fingers, loosen the hide from the muscle, being careful not to tear the meat from the bone.

4. Using your fingers, pull down the hide, forming a tube shape.

5. Using your knife, cut around the anus, leaving the fur and skin there for now.

6. Now, you can easily pull the hide off the rabbit in one piece by gently tugging in a downward movement.

7. You may need to do a few cuts at the front paws to release the hide from the foot joint.

8. Cut off the head to release the hide from the carcass.

9. If you are reusing the hide, put it aside for later.

IMPORTANT REMINDERS

If the blood is not properly drained during slaughter and processing, it may result in an unpleasant taste and odor in the meat. Additionally, excessive blood in the meat could be a sign of inadequate processing techniques. It is crucial to ensure proper slaughtering and processing practices to maintain the quality and safety of rabbit meat.

Removing the rabbit hide before preparing the meat is important for several reasons. First, the skin may carry dirt, bacteria, or parasites that could contaminate the meat if not properly cleaned. Second, the fur itself might be a source of unwanted flavors if left on during cooking.

Last, removing the hide facilitates proper inspection of the meat for any abnormalities, ensuring that only high-quality, safe meat is used for consumption. Overall, it's a hygiene and quality control measure in the preparation process.

STEP THREE — EVISCERATING THE ANIMAL

This step should be performed immediately after removing the hide when the carcass is still hanging from its hind legs.

1. Using a sharp paring knife, sever the bone between the hind legs and under the tail, being careful not to rupture the organs connected to the anus. Using your fingers, twist the organs and the anus out of the way.

2. Cut under the genitals into the belly, slicing in a downward motion past the ribs.

3. Using your fingers, open the rib cage and gently pull out the innards. Take care not to rupture the bladder or the bile sack (the green-colored tube attached to the liver).

4. Use a pair of poultry shears to snip off the hind feet.

5. Now, the rabbit is slaughtered and dressed. Place the carcass in a bucket full of ice-cold water. Add a generous handful of salt to the water, as this will help draw out any excess blood that remains.

After the rabbit has been dispatched, blooded, skinned, and eviscerated, the following step is to divide the carcass into different cuts. The most common cuts are the saddle, legs, shoulders, and ribs. Depending on your market, you can package accordingly. Most restaurants prefer the whole rabbit, but market clients often prefer the rabbit to be already cut, making it easier for home cooking.

Each cut offers a distinctive flavor and texture, making it suitable for different culinary recipes.

Saddle — This is considered the prime rabbit cut as it is tender and versatile. Often, it is used for grilling, roasting, or braising.

Legs — This cut is praised for its lean and flavorful meat. It is ideal for use in stews and casseroles.

Shoulders — This has a slightly tougher texture but is the most flavorful of all the cuts when prepared correctly. It is ideal for slow-cooking methods, such as braising or stewing.

Ribs — Even though this cut is smaller in size, it is jam-packed with flavor. It can be used in various recipes, adding depth and flavor to the dish.

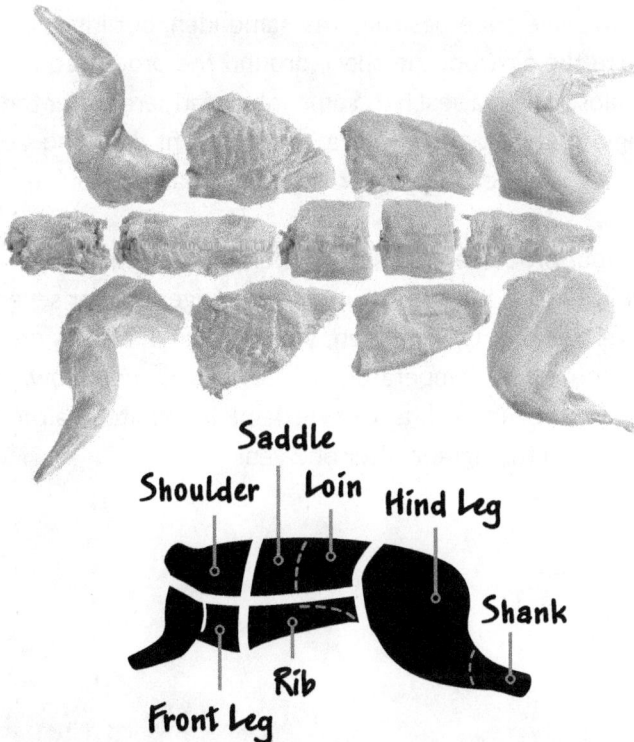

Preservation techniques play an important role in maintaining the freshness and flavor of rabbit meat. The most popular method is freezing, which involves carefully packaging the meat to prevent freezer burn. In the section below, we will discuss more thoroughly how to properly package rabbit meat.

Other preservation methods include canning, curing, or smoking. Canning involves sealing the cooked rabbit meat in an airtight container. Curing and smoking are two of the oldest methods of food preservation. These methods add unique flavor and texture.

Packaging

There are two main ways to package your processed rabbits—vacuum sealing or poultry shrink bags. I recommend both methods, as exposure to air will cause food to spoil faster.

Poultry shrink bags are the easiest to use as you simply place the processed rabbit inside the bag and seal it. The bags are designed to expel excess air when they are dipped in hot water, causing the bag to shrink.

Vacuum sealers are basically the same idea, but instead of water, they remove the air from the space around the processed rabbit, and the bag is closed with a seal bar. Some rabbit farmers prefer the poultry shrink bags method as it is more cost-efficient since no additional equipment is required. However, vacuum sealers provide a neater appearance.

Once your processed rabbit is properly packaged in an oxygen-deficient environment, it can be stored in a refrigerator for seven to ten days before freezing. Once frozen, your packaged rabbits need to be kept at a consistent temperature of 40°F (4°C) or below. Properly packaged and frozen rabbits can be kept indefinitely, although the quality will begin to degrade after one year.

Using a Slaughterhouse

If you are required by state law to use a licensed slaughterhouse or simply prefer to hand off this aspect of the job to someone else, make sure the numbers work for you as far as fees, transportation, travel time, and mileage. If you live far from a slaughterhouse, some states have mobile facilities that will come to your farm for an extra charge.

A reputable slaughterhouse often charges around $5 per rabbit to process, dress, and package. This fee can add up quickly. Typically speaking, if you use a slaughterhouse, processing fees will eat up at least 25% of your overall expenses, which will need to be taken into consideration in your final market price.

Find a slaughterhouse that you trust, as their work will be directly representing your business. If they do good work, then you look good. But if they do sloppy work, you look bad and will lose customers.

How do you find a good slaughterhouse?

- Talk to other farmers.
- Research online.
- Get references.
- Examine their work closely.
- Go and examine the slaughterhouse.
- Talk to the owners.

Before choosing a slaughterhouse, make sure their philosophies about animal welfare are similar to yours. Closely examine the cleanliness of their facilities. If possible, ask to observe how they process, dress, and package a rabbit. Before signing the contract, make sure you know exactly what is included in the full price to avoid unwanted surprises.

When you find the right slaughterhouse to work with, treat them with respect. Be sure to book your appointments on time, and always show up ahead of time, never late. Avoid last-minute cancellations. Bring them a box of doughnuts and coffee once in a while. Working in a slaughterhouse is a tough job that is underappreciated.

ASK THE EXPERTS

What tips or best advice do you have when it comes to the butchering and processing of your rabbits?

Butchering and processing rabbits can be one of the most challenging aspects of rabbit farming, both practically and emotionally. Our experts shared their best tips to help make the process as humane, efficient, and manageable as possible. From choosing the right dispatch method to being well-prepared and having quality tools, the advice below provides valuable insights for making butchering less stressful and more respectful. Learning from mentors and staying mindful of the purpose behind this task can also help new breeders approach butchering with confidence and care.

1. CHOOSE A DISPATCH METHOD THAT WORKS FOR YOU

There are multiple ways to dispatch rabbits, such as cervical dislocation, the broomstick method, a pellet gun, or a choke chain. The best approach is to research and try different methods to determine what works best for the individual, while also ensuring the rabbit is dispatched humanely. Confidence is key to avoiding unnecessary suffering.

"Make sure you are confident. You want to make sure you kill it right and fast, so don't hesitate. I use the broomstick method and it has always worked for me. Be ready for them to twitch a lot after they are dead. It will freak you out a little, like they are still alive."

Amber Irwin
Silver Cloud Rabbitry

2. PREPARATION IS ESSENTIAL

Before butchering, having all tools ready and a well-organized setup is crucial. This includes having sharp knives, a comfortable place to hang and bleed the rabbits, and containers for separating edible parts from waste. Some breeders suggest using specific tools like bone shears or gambrels for convenience. Making sure the area is easy to clean afterward also simplifies the process.

"Don't name them. Butchering is probably the hardest part, but it is necessary. We use two people to butcher. One knocks it out while the other person holds it, and one person skins it. We like to have our table and system set up for a quick, painless, and planned butchering."

Abigail Padilla
Farnash Creek Ranch

3. MENTORSHIP AND LEARNING

Watching online videos, attending workshops, or finding a local mentor are some of the best ways to learn butchering skills. Many breeders mentioned that the first butchering experience can be difficult, and having an experienced person guide the process makes it more manageable. YouTube is a commonly suggested resource for learning different dispatch and butchering techniques.

4. EMOTIONAL CHALLENGES

Butchering rabbits can be emotionally challenging, especially for those who are new to the process or have raised the rabbits since they were kits. Many experts emphasized the importance of remembering the purpose behind the process and suggested not getting too emotionally attached to the rabbits. Approaching the task with respect and mindfulness helps make it a bit easier.

5. QUALITY OF TOOLS AND EFFICIENCY

Investing in quality tools is key to making the butchering process smoother and safer. A good sharp knife is particularly important, as dull blades can lead to accidents. Keeping tools like bone shears, a butcher's apron, and buckets for rinsing on hand makes the process more efficient. Additionally, understanding the anatomy of the rabbit beforehand helps avoid mistakes during processing.

"It's OK to be shaken up when you kill your rabbits. It's hard, especially when you raise them as kits and you watch them grow and take care of them. It's OK to feel remorseful."

Kelly Hurley
Phillips Farmm

"No matter which method you choose, make sure that you have a backup in case it goes wrong or you don't perform it correctly on the first try. When in doubt, double tap. Overkill is ALWAYS more humane than under-kill. I would rather clean up the blood than have the screams of an injured rabbit echoing in my head."

Heather Riddell-Ide
Riddell-Ide Farm

"Invest in good equipment, such as skinning, boning, and butcher knives, carcass hangers, and dispatch tools. Make sure you have a plan on how you are going to set up your area for butchering, that you have enough packaging material and room in the freezer."

Mia Barcenas
Patch of Heaven
Homestead Farm

CHAPTER 11

From Rabbit Farm to Table

Fresh or frozen rabbit meat can be sold year-round. However, getting your rabbit meat from your rabbitry to the consumer's table involves careful planning and smart marketing.

Overcoming Stereotyping

There are a few challenges to marketing rabbits. Probably the biggest obstacle is overcoming stereotypes. It can be a challenge for people to try rabbit meat.

For example, children will devour a plate of chicken wings without even thinking about it, but heaven forbid they think of taking a bite of rabbit meat. If you tell a kid that they are eating rabbit, they will instantly think of the Easter Bunny, Peter Rabbit, Bugs Bunny, or a cute little floppy-eared bunny. Most consumers are taught to believe that meat originates in the back of a supermarket and is always covered in shiny plastic wrap.

Many people assume rabbit tastes gamey and wild, but nothing could be further from the truth. Rabbit meat is fine-grained, tender, and pearly white. A domestic rabbit is no more wild game than a Black Angus steer or an ISA brown chicken.

I highly recommend that you try rabbit meat before you start raising them. You can pick up a fryer/broiler in the frozen food section at any of the larger grocery chains across North America. You may be surprised at the price, as it is considerably more expensive than a whole chicken. Often, the price of rabbit meat is similar to sirloin steak, but prices can vary from state to state. The higher cost has to do with the feed/meat conversion ratio, which we discussed earlier in this book.

Chicken is cheaper than rabbit. Actually, the cost of chicken has never been cheaper, with fowl costing less these days than the same meat did 40 years ago. This is mostly due to the high performance of the American poultry industry. Chicken consumption started to dominate the market when Americans became obsessed with reducing cholesterol, causing the price of chicken to decline and the price of beef to increase.

One way to overcome this stereotyping is to compare rabbit positively to chicken. Rabbit is one of several meats that taste similar to chicken but have a long list of health benefits. Other meats that taste like chicken include frogs' legs, alligator, iguana, turtle, shark, pheasant, and nutria. However, most of us would be wary of trying these exotic animals.

	RABBIT	CHICKEN
Protein percentage	20.8	20
Fat percentage	10	11
Moisture percentage	28	68
Calories per pound	790	810

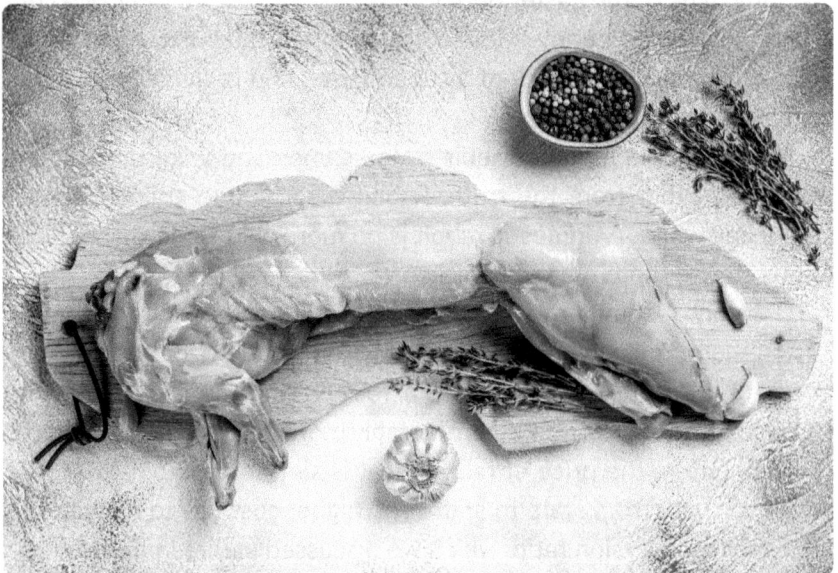

Here are some arguments to persuade consumers to give rabbit meat a try:

- **Easy to digest:** Rabbit is easy to digest. Medical professionals often recommend feeding rabbit to those with weak stomachs or digestive issues. Also, patients on a bland, soft-food diet will welcome the tender texture and mild flavor of stewed rabbit.

- **Faster satiety:** Rabbit has finer bones than chicken and is finer-grained with a chewy texture similar to beef. So, a little bit fills you up faster than chicken. For example, even the United States Navy recognizes this fact by allotting each sailor a six-ounce portion of rabbit meat versus a whopping 12 ounces of chicken.

- **More cost-efficient:** As you will notice in the above table, rabbit contains less moisture than chicken, which means you are not purchasing water but a high-quality protein. Plus, with rabbit you do not pay for the skin, which, in the case of chicken, is fatty, rubbery, and often ends up being thrown in the trash.

- **Extremely versatile:** You can cook rabbit using almost any recipe for chicken and even with some recipes for veal. Your options to prepare rabbit are almost endless as you can prepare it any way you like—fancy or simple. Below, you will find some delicious recipes that are approved by children and adults alike.

- **Real food:** Thankfully, we live in an age where people are more and more concerned about what they are putting into their bodies. They are looking for whole foods that are not overly processed or jam-packed with hormones. Rabbit meat offers just that! With the right marketing and promotion, people can easily be convinced to try rabbit.

Marketing

During the past century, America's history of marketing rabbit has been short compared to beef, chicken, or pork.

Around the turn of the twentieth century, rabbit meat was widely consumed throughout the United States. However, this meat source was not prized as it was considered to be a common meat, often consumed by immigrants and the poor.

During World War II, there was an increase in rabbit consumption due to the fact that the majority of beef was being sent overseas to feed the troops. Thanks to smart marketing and governmental aid, soon, rabbitries were springing up across the country to fill the dietary void with a new protein choice. Within a short period, Americans from all social statuses were eating the new white meat.

Soon, recipes for rabbit were published in newspapers and popular magazines, such as Gourmet. During the war, it was common to see families switch up their Easter ham with an Easter rabbit.

When the war came to an end, beef was readily available. As a result, within a short time, rabbit lost its popularity, and rabbit was soon a rarity in the American diet.

In the 1960s, rabbits returned to Americans' dinner plates when Julia Child made eating this white meat fashionable again. But the fame did not last due to the fact that rabbit was hard to find and more expensive than other meat options.

In 1985, the USDA attempted to bring back the rabbit trend by marketing it in popular newspapers, such as the Washington Post and the New York Times. Once again, there were not enough producers to meet the surge in popularity. However, this history indicates that there is an appetite for rabbit meat if you market your product correctly.

Here are some suggestions to market your rabbits without breaking the bank to pay for expensive advertisements.

Local farmers' markets

Participate in farmers' markets to connect directly with consumers. Display the quality of your rabbit meat and share information about your farming practices.

Online presence

Establish a website or use social media platforms to showcase your rabbit meat. Share engaging content about your farm, your practices, and the benefits of your product.

Community events

Sponsor or participate in community events to create awareness. Offering samples or cooking demonstrations can attract potential customers.

Collaborate with restaurants

Partner with local restaurants or chefs who appreciate high-quality, locally sourced meat. This can introduce your rabbit meat to a wider audience.

Subscription services

Consider offering subscription services where customers receive regular deliveries of your rabbit meat. This can create a consistent customer base.

Educational workshops

Host workshops or webinars to educate people about the benefits of rabbit meat, your farming practices, and cooking techniques. This positions you as an expert in your field.

Customer testimonials

Encourage satisfied customers to share their experiences through testimonials. Positive reviews can build trust and attract new customers.

Quality packaging

Invest in professional and appealing packaging. Eye-catching labels and information about your rabbit meat's quality can influence purchasing decisions.

Local specialty stores

Approach specialty food stores or butchers who may be interested in carrying your rabbit meat, especially if it aligns with their focus on local and high-quality products.

Word of mouth

Leverage satisfied customers as ambassadors for your rabbit meat. Positive word of mouth is a powerful marketing tool.

Remember to comply with local regulations and use a combination of these methods to create a comprehensive and effective marketing strategy for your rabbit meat.

Labeling

Clients are accustomed to seeing specific wording or labeling when purchasing meat. Knowing, understanding, and using common label terms can help your client to find exactly what they are looking for. So, make it easier for your customers by using these common expressions on your rabbit meat packaging.

Fryer or young rabbit

This expression refers to a young rabbit, often between 12 to 16 weeks of age, depending on the breed. Typically, these rabbits weigh between one and a half to three and a half pounds after being processed. A fryer or a young rabbit is the most tender rabbit meat due to its fine grain. Generally, a fryer can be used for most conventional chicken recipes.

Roaster or mature rabbit

These terms refer to an older adult rabbit that weighs more than four pounds dressed. Often, for a rabbit to weigh more than four pounds before processing, it will be eight months or older. A roaster has a firmer and tougher texture than a young fryer due to the coarser grain. Despite their name, roasters are mostly used in braising or stewing techniques.

Giblets

As with poultry, giblets in rabbits refer to the liver, heart, and kidneys. They are often sold separately from the rest of the rabbit carcass. Giblets can be lightly sauteed, fried, or used in paellas, pates, or terrines.

Head on

This expression is pretty much self-explanatory, as it is a dressed rabbit with its head still attached. Often, restaurants will order rabbits with the head on for more rustic culinary traditions.

Labels should be made of a waterproof material that will not slide off the plastic packaging. Also, make sure the ink in your printer is suitable for the labels, as some inks can smear or smudge from excess condensation. Commercial labels can be purchased at most farmer supply stores or online.

FINE QUALITY ORGANIC
Rabbit

Elevate your culinary experience with our high-protein rabbit meat. Taste the difference that comes from ethical farming practices and a commitment to excellence.

ESTD ★ 1998
LETTERBOX FARM

NATURAL *Farm* PRODUCT

LOCAL PURVEYORS OF *Fresh* **MEAT**

850 g (30 oz) | KEEP REFRIGERATED | Best Before: Jun 20, 2025

WHAT OTHER COMMON INFORMATION DO YOU NEED TO INCLUDE ON THE LABEL?

Requirements for dressed rabbits for commercial sale vary from state to state and county to county. Here is a short list of the most common regulations that should be included on the label.

Statement of identity

This refers to the name of the rabbitry or the place of business of the responsible party. For example, if your rabbitry is called Letterbox Farm, that business name should be included on the label. It is optional to add your address and phone number; however, it is free marketing for future business.

Net quantity

Within the US, the prepackaged product must declare the weight using both metric and imperial units. For example, 500 g net quantity / 17.6 oz net quantity.

Date markings

The majority of meat products, when packaged and stored correctly, have a durable life of 90 days. The expression must say Best Before and then be followed by the durable life date. For example, Best Before: Feb 24, 2024.

For all meat products, including rabbit, that will be sold by a retailer or commercial store, the original packaging should have the date the product was processed and packaged and the durable life date.

Storage instructions

Storage instructions must be shown on the principal label using one of the following expressions, whichever is applicable: "Keep Refrigerated" or "Keep Frozen."

Nutrition labeling

Rabbit meat is exempt from nutrition labeling, as it is a single raw ingredient. However, for marketing purposes, you may decide to add the information to your label.

Lot code

This requirement varies from state to state, so be sure to check local regulations regarding using a lot code. A lot code is a unique identifier for traceability purposes.

Pricing

When pricing your rabbit meat on the label, be sure to clearly display the price per unit according to weight. Specify the weight of the rabbit per pound, the actual weight of the rabbit, and the corresponding price.

Ensure that your pricing information is clear, accurate, and complies with local regulations. If you're unsure about specific requirements, consider consulting with local health departments or agricultural authorities to ensure your labeling meets all necessary standards.

How to Price Your Rabbits

When pricing rabbit meat, consider factors like production costs, market demand, and local competition. Account for expenses such as feed, housing, and veterinary care. Research current market prices for rabbit meat in your area and adjust your pricing accordingly. Building relationships with local markets or restaurants can also help you understand the demand and set a competitive yet profitable price.

WHOLESALE AND COMMERCIAL

For wholesale and commercial pricing of rabbit meat, it's crucial to factor in bulk quantities and potential long-term partnerships. Consider production scalability and distribution costs, and negotiate with buyers based on volume. Research competitors' pricing strategies in the wholesale rabbit meat market and aim for a competitive yet profitable rate. Building strong relationships with potential commercial buyers can lead to consistent sales and partnerships.

RETAIL

For retail pricing of rabbit meat, assess your production costs, including feed, processing, and packaging. To stay competitive, research local market prices for rabbit meat. Consider positioning your product based on quality, freshness, or any unique selling points. Keep in mind consumer preferences and purchasing power in your target market. Regularly reassess and adjust your retail prices to stay competitive and meet market demands.

There are a few other aspects to take into consideration when pricing your rabbits, such as transportation, packaging, and more.

TRANSPORTATION

Transportation significantly influences the price of rabbit meat. Factors such as distance, fuel costs, and logistics impact transportation expenses. Longer distances and complex routes may increase costs, affecting the overall price. Efficient transport management and negotiating favorable rates with carriers can help mitigate these expenses. Consider optimizing your supply chain to minimize transportation costs and maintain competitive pricing in the market.

PACKAGING

Packaging fees can vary based on the type of materials you choose and the scale of your operation. Costs may include the price of boxes, labels, and any additional branding elements. To determine packaging fees, consider the quantity and quality of packaging needed for your rabbit meat. It's also advisable to explore cost-effective packaging solutions without compromising on safety and presentation. Negotiating bulk rates with suppliers and optimizing your packaging process can help manage these fees more efficiently.

FARMERS' MARKET FEES

Fees for participating in farmers markets vary widely and depend on the specific market's policies. Some markets charge a flat fee for a booth space, while others may take a percentage of your sales as a commission. It's essential to inquire directly with the organizer of the farmers' market you're interested in to understand their fee structure. Additionally, consider the potential benefits in terms of exposure and direct consumer interaction when evaluating the cost of participating in a farmer's market.

MISCELLANEOUS COSTS

In addition to packaging and transportation, other fees involved in retailing rabbit meat may include:

1. **Processing fees:** These are costs associated with slaughtering, butchering, and preparing the rabbit meat for sale.
2. **Marketing and advertising costs:** These are expenses related to promoting your rabbit meat products, whether through online channels, local advertising, or promotional materials.
3. **Permits and licenses:** You must ensure you are in compliance with local health and food safety regulations, which may involve obtaining permits and licenses, each with associated fees.
4. **Storage costs:** If you need cold storage for your rabbit meat, consider expenses related to refrigeration or freezer facilities.
5. **Insurance:** Protect your business with insurance coverage, which may include liability insurance, product liability insurance, or coverage for property and equipment.

6. **Market membership fees:** Some retail markets or platforms may require membership or participation fees.

It's crucial to thoroughly research and budget for these various costs to ensure a clear understanding of the financial aspects of retailing rabbits.

How Are Rabbits Inspected?

According to the Federal Meat Inspection Act (FMIA), the USDA authorizes Food Safety and Inspection Services (FSIS) to inspect swine, cattle, sheep, and goats. Under the Poultry Products Inspection Act (PPIA), the FSIS is authorized to inspect domesticated poultry, which includes chickens, turkeys, ducks, geese, guineas, ratites, and squabs.

However, there seems to be a little loophole when it comes to rabbits. Congress has not established the inspection of rabbits under either the FMIA or the PPIA; therefore, the inspection of rabbits is not required and is voluntary. According to the Agricultural Marketing Act, voluntary inspection can be made of the following animals: buffalo, antelope, reindeer, elk, deer, migratory waterfowl, game birds, and rabbits.

Voluntary inspection means each rabbit and its internal organs have been inspected for any signs of disease. Depending on state laws, the inspection is often performed by a licensed state inspector. This inspector may charge up to $100 per hour. This makes the whole process financially prohibitive for almost all medium to small scale producers.

HOW ARE RABBITS GRADED?

The reality is that grading rabbit meat, unlike other livestock, is almost never done. Since inspection of rabbits isn't required and would be cost-prohibitive for most producers, grading standards are not something you'll need to worry about at all.

While USDA guidelines technically allow for rabbit meat to be graded if it has been inspected, the USDA's Agricultural Marketing Services does not typically grade rabbits. They focus on products like eggs, poultry, and dairy, leaving rabbit grading as a rare practice.

HORMONES AND ANTIBIOTICS

Yes, you can label your meat rabbits as hormone- and antibiotic-free if you follow appropriate practices. To make this claim, ensure that your rabbits have not been treated with hormones or antibiotics throughout their lives. Keep detailed records of their diet and medical history to support your claim in case of inspection.

Labeling meat as hormone- and antibiotic-free can appeal to consumers who prioritize natural and healthy food options. It emphasizes transparency and may attract those seeking products with fewer additives. However, it's crucial to adhere to labeling regulations to maintain trust with consumers and avoid legal issues. Verify with local authorities or agricultural agencies for specific guidelines in your region.

Additional Information for the Consumer

Providing customers with safe storage and thawing instructions is a valuable aspect of marketing your rabbits. It demonstrates your commitment to their satisfaction and ensures the best possible eating experience.

1. **Customer satisfaction** — Clear instructions help customers store and handle the rabbit meat properly, ensuring its quality and safety. Satisfied customers are more likely to become repeat buyers and recommend your products.

2. **Quality assurance** — Proper storage and thawing guidelines contribute to maintaining the quality of your rabbit meat. This reinforces the message that you prioritize the well-being and satisfaction of your customers.

3. **Brand reputation** — Offering useful information establishes your brand as reliable and customer-focused. This positive reputation can contribute to building trust and credibility in the market.

4. **Reducing wastage** — Providing guidance on safe storage and thawing helps customers avoid mishandling the product, reducing the likelihood of spoilage and waste. This aligns with sustainability principles and can be an attractive feature for environmentally conscious consumers.

5. **Educational value** — Sharing storage and thawing recommendations can be an educational opportunity. It showcases your expertise in rabbit meat and demonstrates your commitment to consumer education.

Incorporating these instructions into your marketing materials, such as packaging or informational inserts, adds value to your product and enhances the overall customer experience.

SAFE STORAGE TIMES

Upon purchasing, immediately take the rabbit home and place it in the refrigerator at 40°F or below. Use within two days or freeze at 0°F. If kept frozen continuously, the rabbit will be safe to eat indefinitely; however, the quality will begin to diminish. For best quality, use frozen rabbit within a year and any pieces within nine months. It is safe to freeze the rabbit in the original packaging.

SAFE THAWING

There are three ways to safely defrost a rabbit: in the refrigerator overnight, in cold water, or in a microwave on the defrost setting. Never defrost a rabbit at room temperature.

- **Refrigerator:** This method is ideal as it slowly and safely unthaws the rabbit in a cold environment. Rabbits with the bone in may take up to 24 hours to completely unthaw in the refrigerator. Once unthawed, you may store it in the fridge for two days before cooking.
- **Cold water:** Leave the rabbit in the original airtight packaging. Submerge the rabbit in cold water, changing the water every 30 minutes so that it continues to thaw at a safe temperature. A whole rabbit can take up to two to three hours to unthaw using this method. Plan to cook the rabbit immediately after thawing using the cold-water method.
- **Microwave:** Use the defrost setting on your microwave. Plan to cook the rabbit immediately after thawing using this method, as some parts may become warm and start to cook.

SAFE COOKING

To safely cook rabbit, ensure it reaches an internal temperature of at least 165°F (74°C) to kill any harmful bacteria. Use a meat thermometer to check. Additionally, handle raw rabbit with proper hygiene, separate it from other foods, and cook it within a day or two of purchase for freshness.

A whole two-pound rabbit should take about 60 to 90 minutes to roast. Stuffing it will add an additional 30 minutes to the cooking time.

SAFE HANDLING OF LEFTOVERS

- Place the leftovers in the refrigerator within two hours after cooking. Use any leftovers within three to four days or freeze.
- Use frozen cooked rabbit within four to six months for best quality.
- Reheat leftovers to 165°F.

Recipes

Here are some of my friends and family's favorite recipes using rabbit meat. Trust me, you will love them too!

OVEN-FRIED RABBIT

If you love "fried" chicken but without all the grease and extra calories, then this recipe will be your favorite.

SERVES 4

1 fryer-broiler rabbit (2 to 2 ½ pounds dressed), cut into pieces

2 eggs, mixed together with a fork

3 cups salted potato chips, finely crushed

¼ cup butter, melted

DIRECTIONS

1. Dip the rabbit pieces in egg, then coat with the crushed potato chips evenly.
2. Preheat the oven to 375°F (190°C).
3. Place the melted butter in a pan and arrange the coated rabbit pieces on top.
4. Bake in the preheated oven for 30 minutes.
5. After 30 minutes, flip the rabbit pieces and bake for another 30 minutes. Test the pieces with a fork. The meat should be tender, juicy, and well cooked.

RABBIT CACCIATORE (HUNTER'S RABBIT)

Originally, chicken cacciatore was prepared by Italian hunters using wild game such as rabbits and squirrels. Over the decades, it has been adapted for a more Western palate and made with chicken. However, beware: once you try rabbit cacciatore, there will be no going back to chicken.

SERVES 4

1 fryer-broiler rabbit (2 to 2 ½ pounds dressed), cut into pieces

1 large onion, chopped

1 can tomato paste

3 garlic cloves, minced

2 bay leaves

Oregano, to taste

Salt and pepper, to taste

DIRECTIONS

1. Precook the rabbit pieces in salted, boiling water and simmer until tender (approx. 20 to 30 minutes). Remove the pieces of cooked rabbit and set the broth aside.

2. Add the onion, tomato paste, oregano, bay leaves, garlic, salt, and pepper to the rabbit broth and let simmer on low heat for one hour.

3. Put the precooked rabbit pieces back into the reduced broth and simmer together for five minutes. Serve on top of spaghetti.

CIDER-BRAISED RABBIT ROAST

Slow-cooked rabbit roast with caramelized onions and crispy sage is the easiest dish you can make to cozy up to in the winter months.

SERVES 6

2 fryer-broiler rabbits (2 to 2 ½ pounds dressed), cut into pieces

Kosher salt and pepper, to taste

¾ tbsp flour

Pinch of garlic and celery salt

Olive oil, as needed

3 yellow onions

4 shallots, halved

1 cup apple cider

1 cup dry white wine

2 tbsp chopped fresh thyme

DIRECTIONS

1. Season each piece of rabbit with salt and pepper. Set aside.
2. In a plastic bag, combine the flour, garlic, and celery salt. Mix well.
3. Add the seasoned pieces of rabbit meat to the bag and shake until all the pieces are well-coated in the flour mixture.
4. Heat the olive oil in a skillet.
5. Brown the rabbit pieces in the olive oil, flipping as needed to ensure all sides are well browned.
6. Add the onions and apple cider, white wine, and thyme. Simmer until the onions are translucent and the rabbit is tender. This should take about one hour or more if the pieces are on the larger size. Test with a fork for tenderness.
7. Serve with mashed potatoes.

Glossary

Every industry, whether related to livestock, apiaries, or rabbitries, has its own unique jargon. If you are new to rabbit farming, these terms below will help you sound like a veteran in no time.

Agrarian — Related to the culture of a farming lifestyle.

Buck — A male rabbit, often used for breeding purposes.

Colony — A group of rabbits, also called a herd.

Cull — To selectively euthanize or slaughter an animal early, often to improve the future quality of breeding stock.

Cuniculture — The agricultural practice of breeding and raising domestic rabbits for their meat, fur, or wool.

Dam — A female rabbit that is a mother, often referenced when discussing lineage.

Doe — A female rabbit, often used for breeding purposes.

Dress weight — The weight of the rabbit's carcass that has been skinned and cleaned.

Forage — To wander or go search for food.

Fryer — A young, tender rabbit that is less than 14 weeks old and weighs less than four pounds dressed.

Grand Dam — The maternal grandmother of a rabbit, important in tracking lineage.

Grand Sire — The paternal grandfather of a rabbit, used in lineage tracking.

Grow-out or junior — A young rabbit that has not reached maturity yet.

Herbivore — An animal that feeds on plants.

Herd — A group of rabbits, also called a colony.

Kindle — When a rabbit gives birth.

Kit or kitten — A baby rabbit.

Live weight — The weight of the rabbit while it is still alive; often, the animal is weighed prior to being processed.

Process — To slaughter or kill an animal for food at a scheduled time.

Roaster — A mature rabbit of any weight.

Ruminant — An even-toed mammal that chews the cud regurgitated from its rumen. This includes cattle, sheep, antelopes, and deer.

Sire — A male rabbit that is a father, also used in lineage discussions.

Stewer — An old and large rabbit whose meat is tough and, therefore, is used in meat stews.

Tractor — A moveable pen with an open bottom used for letting rabbits forage on grass.

Weanling — A six- to eight-week-old rabbit that no longer relies on milk and can be separated from its mother.

www.ingramcontent.com/pod-product-compliance
Lightning Source LLC
Chambersburg PA
CBHW062126020426
42335CB00013B/1121